THE MAN WHO STOPPED TIME

THE MAN WHO STOPPED TIME

THE ILLUMINATING STORY OF EADWEARD MUYBRIDGE— PIONEER PHOTOGRAPHER, FATHER OF THE MOTION PICTURE, MURDERER

BRIAN CLEGG

Joseph Henry Press
Washington, D.C.

Joseph Henry Press • 500 Fifth Street, NW • Washington, DC 20001

The Joseph Henry Press, an imprint of the National Academies Press, was created with the goal of making books on science, technology, and health more widely available to professionals and the public. Joseph Henry was one of the founders of the National Academy of Sciences and a leader in early American science.

Library of Congress Cataloging-in-Publication Data

Clegg, Brian.
 The man who stopped time : the illuminating story of Eadweard Muybridge: father of the motion picture, pioneer of photography, murderer / Brian Clegg.
     p. cm.
 Includes bibliographical references and index.
 ISBN-13: 978-0-309-10112-7 (cloth : alk. paper)
 ISBN-10: 0-309-10112-3 (cloth : alk. paper) 1. Muybridge, Eadweard, 1830-1904.
 2. Cinematographers—United States—Biography. 3. Photographers—United
States—Biography. I. Title.
 TR849.M84C64 2007
 778.5'3092—dc22
 [B]
                        2006032372

Cover credit: Eadweard Muybridge's Motion of a Galloping Horse © CORBIS

Printed in the United States of America

For Gillian, Chelsea, and Rebecca

CONTENTS

PREFACE

Eadweard Muybridge was a Victorian marvel. Born in Kingston-upon-Thames, England, in 1830, he was to venture across the Atlantic and traverse the American continent, settling in the still emerging and sometimes dangerous environment of gold rush California. Although his photographic images remain iconic, the man himself is practically unknown. Yet he has three powerful claims on our attention.

First, Muybridge was a superb, larger-than-life character. His story is not just one of technological breakthrough and artistic endeavor but of a very human struggle. His personal life teetered on the edge of disaster. He survived a near fatal stagecoach accident, only to come even closer to death in a trial where a guily verdict would have seen him hanged. Even his business dealings seemed incapable of remaining straightforward for long.

Second, he was an artist. His photographic work was highly popular in his day and remains striking still. In pure photographic terms, his pictures of Yosemite and his huge panorama of San Francisco are remarkable. At a time when photography was still an arcane art, Muybridge was a master.

Finally, and most importantly, he bridged the gap between art, science, and technology with his thousands of sequences of motion photographs. It is no surprise that in his European tour his lectures sold out at both the Royal Institution and the Royal Academy—respectively among the United Kingdom's top scientific and artistic venues. His work on the photography of motion was groundbreaking. And there was something more still.

Just as Charles Babbage has a rightful claim to be the first in the field of computing, even though his never-constructed mechanical Analytical Engine was not the direct ancestor of the modern PC, so Muybridge brought moving pictures to life before anyone else, even constructing the world's first purpose-built movie theater. His technology was a dead end, just like Babbage's, but he was the conceptual father of cinema and TV. That he does not have the same wide regard that Babbage has in his field is largely due to a determined attempt to remove Muybridge from his place in history. But Muybridge, and his achievements, will not be forgotten.

PART ONE

SHADOW OF THE NOOSE

ONE

DEATH AT YELLOW JACKET

Was not the information Muybridge received on the morning of the homicide sufficient to impel him to insanity and the rash act of homicide? All the evidence shows the holy and tender regard which he had for his wife. The marriage relation is ordained by God; it is not merely a civil contract. . . . God hath said "Let no man commit adultery" and "Thou shalt not covet thy neighbor's wife"; on these two Commandments rests all the law. This law was declared more than 3,000 years ago, and to-day we still revere it. "The man that committeth adultery with another man's wife, even he that committeth adultery, shall surely be put to death. So shalt thou put away evil from Israel." Though this be not the law of California, I believe it is a safe moral guide.

Cameron King speaking for the defense in the Muybridge trial, quoted in *The Napa Daily Register*, February 6, 1875

At four minutes to four on the afternoon of October 17, 1874, pedestrians scattered as a tall, wild-eyed man with a flowing Tennysonian beard burst out of the studios of San Francisco photographers Bradley and Rulofson. There were just four minutes to go before the ferry for Vallejo departed from the dock, the best part of a mile away. The man, Eadweard Muybridge, had little chance of making the boat, yet it was the only way to reach Calistoga in the Napa Valley in time for an appointment with murder.

Though 44 years of age—and looking even older—Muybridge had kept fit. He ran down the steep streets all the way to the waterfront. The ferry was late. As he pounded onto the dock it was still taking its final passengers onboard. Muybridge was on his way to find Harry Larkyns. His wife's lover. The father of their child.

The autumn morning mist had entirely cleared away by then; it was a dry, cold afternoon. The slow boat worked its way up the straits, past Angel Island, through the wide expanse of San Pablo Bay and into the Carquinez Strait, taking around two hours to reach the small port of Vallejo. Normally the photographer in Eadweard would have been intrigued by the constantly changing landscape—now, though, he sat stiff and cold, staring straight ahead.

After a short walk from the landing he transferred to the newly built railroad. It had been opened only three years before and covered the 40 miles up the green heart of the Napa Valley, through the townships of Napa and St. Helena, to reach Calistoga. Although the wood-paneled railroad coaches were the latest technology, laboriously transferred from a distant East Coast factory, they had no heating. The journey was unremittingly wintry.

Outside, in the fading light, the beautiful rolling pastures of the lower Napa Valley gave way to more rugged hills. The view, so reminiscent of Italy or southern France, passed by without attracting Eadweard's attention. When the train reached the small terminus at Calistoga, little more than a wooden shack and a pile of logs to stop runaway trains, he asked for directions to the livery stable, then headed straight through the growing darkness toward the far end of town.

To the stable hand, perhaps drying a steaming horse that had just arrived from a long journey, Eadweard must have seemed a dramatic figure. He swept in from the cold as if the weather did not exist, the image of a patriarch out of the Bible with his long beard and piercing eyes. Eadweard asked for a rig with a driver to take him up the valley to Pine Flat. The liveryman was reluctant. The night was pitch black by now, and bitterly cold. The roads were still thick with mud from the recent rains. Why, he wondered, would his customer want to go out that far on such a night?

Eadweard told him he had urgent business with an acquaintance, Major Harry Larkyns. It was Larkyns that he was following out to Pine Flat. The liveryman was relieved. He told Eadweard that his quarry was much closer than he thought, just over an hour's drive away at the Yellow Jacket ranch. After a little bargaining, the liveryman agreed to let Eadweard take a rig and his stable boy, George Wolfe, to drive it.

Muybridge accepted. As they left the stables, he offered the boy a $5 bonus if he made the trip safely and in good time.

Wolfe was a sociable young man and for the first part of the journey he indulged in idle conversation about the landscape with his passenger. He would later say that Muybridge seemed quite normal—genuinely interested in their surroundings and the boy's tales. About an hour into the journey, though, Eadweard grew uncomfortable. He took out his revolver, handling it nervously. He asked Wolfe if he had ever been stopped on the road by robbers. The boy shrugged. No, he never had been attacked. Eadweard commented that Wolfe was lucky: he, Muybridge, had been.

The photographer looked down at his gun. He was worried, he said, that his pistol might be fouled. Would the horses react badly if he fired off a round? Wolfe, excited by this strange man, said he thought the horses would be just fine. Eadweard pulled back the loading lever and fired off a shot into the night, sending wildlife flying, the sound of the gunshot echoing around the hills for long seconds afterward. The horses' ears twitched, but they plodded on as if nothing had happened. The conversation between Wolfe and Eadweard never really restarted; they had reached a new stage in the journey. The two travelers drove on to the northwest, climbing 1,000 feet up the jagged ridge separating Napa and Sonoma counties. The rig's lamps threw stark shadows over the bumpy track that seemed to disappear into nothing only feet in front of them.

Before long, they passed the lights of Foss Station and then came close to Knight's Valley ranch house, the nearest neighbor to the Yellow Jacket. Eadweard told Wolfe to stop. He climbed down and rapped on the rough unpainted wood of the ranch house door. The building itself was lost in the darkness. When the door opened, Eadweard asked the silhouetted figure of a man in the doorway if he had heard that Larkyns was up at Yellow Jacket. The answer, ringing out of the darkness, was a "yes." Eadweard thanked his informant, climbing back on the rig without a word to Wolfe. The boy drove on.

Two miles later, near the top of a steep slope, the Yellow Jacket ranch house came into sight. Despite the impressive-sounding title it was only a small white-painted wooden cottage, the home of William

Stewart, the superintendent of the Yellow Jacket mine. As they drew up outside, Eadweard asked Wolfe to wait. He stepped down and crossed to the house. In front of him, a narrow wooden veranda led to the front door.

Eadweard rapped heavily on the door. A young man he didn't recognize, Benjamin Prickett, came to answer the knock and Eadweard told him that he needed a brief word with Major Harry Larkyns. He was told that Larkyns was in the parlor, playing cribbage with some women friends. Would Muybridge care to step inside? Eadweard said he would rather not go in. He stood in the shadows and asked for Larkyns to be sent out.

It had been a good day for Harry Larkyns. He was in pleasant company and luck had smiled on him. He had drunk perhaps one glass of wine too many, but he was still capable of holding a good hand of cards. When he heard that someone was waiting for him, he had no idea who it might be, but stood up from the table and took the few steps through the kitchen to the doorway beyond.

"I can't see you," Larkyns said, peering into the shadows. He could see the glowing halos of the lamps on the rig, but they did not illuminate his visitor, and the lights from the house were too dim to show any detail outside. A voice from the doorway, a very quiet voice only just audible over the chatter from the table behind in the parlor, asked him to come closer. Larkyns took a few more steps, peering into the darkness.

For a moment Eadweard just looked at him. Looked at this handsome swindler, Harry Larkyns, the destroyer of his marriage. When he spoke again, it was as if it were someone else's voice, someone far away.

"My name is Muybridge," he said, "and I have a message for you from my wife."

Before Larkyns could take in what had been said, Eadweard pulled the trigger on his ready-cocked Smith and Wesson revolver.

The photography of motion that Eadweard Muybridge pioneered would eventually reveal the elegant dance of a bullet in flight. When fired from a rifle, a bullet is set spinning by the grooves that line the barrel. It is held as close as possible to a straight line by the gyroscopic effect of this spin, though nothing can prevent the relentless pull of gravity eventually bringing it to earth. A pistol bullet is more easily buffeted by the chaotic influence of air resistance, making it less accurate over a distance. At a range of 5 feet, though, there is little chance of missing. Traveling at around 600 miles an hour, the 0.32-inch rimfire bullet from a Smith and Wesson number two revolver traverses such a distance in five thousandths of a second.

The bullet slammed home in Harry Larkyns's chest.

He let out an anguished yell. One eyewitness claimed that with this cry, like that of a wounded beast, Larkyns had rushed straight through the house—bleeding copiously—and out the back door. The single-story wooden structure had been built right up against a huge oak tree, its great trunk buttressing the side of the building. More likely though, as Muybridge had shot him through the heart, Larkyns would not have made it through the house. He was probably already by the tree when he was shot. He tried to cling onto the rough bark of the tree, but could not get a grip and sank to the damp earth. Within seconds he was dead.

Eadweard made no attempt to escape. He strolled into the parlor as if on a social visit and apologized to the ladies, saying, "I am sorry this little trouble occurred in your presence"; then he waited for someone else to take charge. Stewart, the mine superintendent, made a citizen's arrest. The same rig that had brought Eadweard out to the ranch took him back to Calistoga, where he was held at the Magnolia Hotel, the hotel Larkyns had stayed in when he first came out that way, until he could be taken on to the Napa County Jail the following Monday. Without protest, without comment, Eadweard was locked up. He sat in his cell, quiet, composed, seeming to accept his fate.

The county sheriff had no doubt at all what Muybridge had coming to him. This man had clearly committed murder; there was no possibility it could be anything else. He had witnesses to Muybridge's single-minded arrival in Calistoga; he had the word of Wolfe the rig driver and the testimony of the shocked inhabitants of the Yellow Jacket ranch house. He had the corpse of Major Larkyns. And he had the gun. The simple entry in the sheriff's office at Napa reads:

Muybridge, E. S. - Murder - Arresting Officer Cromwell - arrested September 19th.

Muybridge was, without doubt, destined for the gallows.

TWO

A BLOW TO THE HEAD

We here present to the Public, the first San Francisco Directory yet published.

In an ordinary work of the nature, in a long-established community, the difficulties are such as to require the closest care. . . . All these disadvantages prevail here in a greater degree, in addition to many others peculiar to this place alone. The very frequent change of business residence, rendered necessary by the almost daily change in all the thoroughfares of trade in the City—the great variety of inhabitants, comprising denizens of every clime, creed and country—their very mixed occupations, (with the minor difficulties of foreign languages and names,) all present obstacles to a perfect comprehensive summary.

Introduction to the first *San Francisco City Directory,*
published September 1852

Muybridge was to spend nearly four months in the confines of Napa County Jail, waiting for his trial for murder and a potential death sentence. A diet of Wild West movies makes it easy to imagine the county jail as a shack with a barred cell in the corner, but this picture would not be accurate in the case of the Napa County Jail. The building housing the jail was also the courthouse, and the architect had produced a cross between an exotic eastern palace and a Victorian railroad station, adding an onion-domed tower reminiscent of a compact model of the upper parts of the Taj Mahal to the simple straight lines of the main building.

Muybridge was able to make his living conditions a little more comfortable at his own expense. As was common at the time, he paid to have his meals sent in—in this case, from a Napa hotel nearby. He was allowed to buy and bring in writing materials and was given access to books. Yet this was still four months of incarceration with a death sentence hanging over him. Four months to contemplate how he had reached this state. To think not only of the terrible events that had put him in jail but also of his life before emigrating from the very different surroundings of his birthplace, Kingston-upon-Thames in England.

Kingston was, and still is, a combination of refinement and commerce, a suburb of London with a distinctive character, a borough with a long-reaching history that has probably been occupied for around 5,000 years, but came to prominence only when the Saxons took control of this part of England around the sixth century. The kings of Wessex had a palace there, and there the name originated as "king's town." By 838, Kingston was considered important enough for King Egbert of Wessex, grandfather of Alfred the Great, to hold his Great Council there, in a location described in contemporary records as "that famous place called Cyningestun." Seven Saxon kings, ending with Ethelred II (the Unready) were crowned there, which led many accounts to suggest that Kingston means the place where kings are made—a nice image, but untrue.

These events, far back in history, were Kingston's great claim to fame. The royal borough was a faded remembrance of great things forever afterward. By 1830, commerce had triumphed, replacing any grandiose royal connections as Kingston's legacy. And to John and Susannah Muggeridge, dealers in corn and coal, it provided both a home and a workplace. Muggeridge's mother was originally called Smith, making her Susannah Smith, as was her mother before her.

The Muggeridges lived over the shop at 30 West-by-Thames Street (now more prosaically 30 High Street), and it was there, on April 9, 1830, that their son Edward James (later to become Eadweard Muybridge) was born. He came three years after his brother, John, and was joined later by George and Thomas with a rare regularity in 1833 and 1836, respectively.

The family premises backed onto the river Thames, hence their expansion from their established business as corn merchants to that of selling coal to the barges that tied up on the wharf behind the build-

Muybridge's birthplace in Kingston-upon-Thames, England (above the shop).

ings. Nowadays two blocks of fashionable apartments stand in between the river and 30 High Street, but back then the Muggeridge's yard ran straight to the embankment. West-by-Thames Street was a busy mix of industry and homes, with the beer trade also well represented in the area—malt houses, a brewery, and a pub called The Ram were among near neighbors.

The setting of the Muggeridge home does not sound hugely attractive—though it was near the river, the Thames at this point is more industrial than picturesque—but the upper floors of the house form a surprisingly attractive late Georgian building (now faded and used as the offices of a design consultancy, over a computer store) and the family lived over the shop fairly comfortably: a paneled, eminently Victorian household of solid, respectable lower-middle-class steadiness.

Even so, there was something unusual about the Muggeridges. Eadweard was not the only brother to change his surname; his elder brother, John, who was training to become a doctor, became John Wybridge before his death at the age of 20. John's decision to change his name may have influenced Muybridge's later modification of his own name. Muybridge's father, also John, had died when Eadweard was only 13, leaving his mother and his elder brother as the greatest influences in his life.

One contemporary account of the young Muybridge shows him to be already very inventive. A young cousin, Maybanke Susannah Anderson, later wrote a memoir of her early life in which she described Muybridge as

> an eccentric boy, rather mischievous, always doing something or saying something unusual, or inventing a new toy, or a fresh trick.
>
> To a quiet little girl, much younger than himself, he was a sort of hero. I was always willing to listen, or to render small services to Edward, fetching and carrying for him, even though I sometimes had a misgiving as to the wisdom of the business.

The family environment may have been loving, but Muybridge found it stifling, wanting something more than could be offered by this female-dominated world and the safe but dull environment of Kingston-upon-Thames. Another of Muybridge's cousins, Edward Smith, had already emigrated to New York, where he sold stationery, toys, and musical instruments. Very likely, word of the New World's liveliness and energy had reached Muybridge from cousin Edward. Exactly when he crossed the Atlantic is not known, but Eadweard was to follow his cousin to America.

Muybridge never recorded any detail of his journey to America until he was in his retirement, and even painstaking research by academic Robert Bartlett Haas has never uncovered a record of Muybridge's crossing in the voyage reports of the time. When he was in his seventies, Muybridge said that he had crossed the Atlantic in 1852, and there is no reason to doubt his memory or his honesty.

By then, Muybridge had taken the first step on his transformation from simple Edward Muggeridge to the more enigmatic Eadweard Muybridge. In 1850, a year or two before his crossing to New York, Kingston was celebrating. An ancient stone, thought to be the coronation stone of the Saxon kings crowned in Kingston, had been unearthed from the ruins of the Saxon chapel of St. Mary. Of itself, the stone doesn't look too thrilling—a rough pillar less than a yard high made of hard silicified sandstone, the same material as the sarsen stones used to build Stonehenge. It was initially used unceremoniously as a mounting block for scrambling up onto horses in the Market Place. But in 1850 it was re-erected and set in a celebratory plinth, in the road in front of the Guildhall.

The stone was later moved a few yards to its present site outside the rebuilt guildhall on a concrete raft over the Hogsmill River. On the 1850 plinth, which is adorned with coins from each of the sovereigns' reigns, we learn that among the kings crowned on this stone were two

The King's Stone, Kingston-upon-Thames, England.

Edwards. (Exactly how they were crowned "on" a pillar isn't clear—perhaps they sat on top of it, or it was built into a seat like the current British coronation stone, the Stone of Destiny, or the Stone of Scone.) The plinth uses the Anglo-Saxon form of the name Edward, Eadweard, and as Edward Muggeridge was known as Eadweard Muggeridge from around 1850, and the stone was located within sight of the upper windows of the Muggeridge home, it seems very likely that this inscription was the inspiration for the first stage of his transformation into Eadweard Muybridge.

Like his cousin Edward before him, Muybridge selected New York as the starting point of his adventure abroad. There is no evidence, at this point in his life, of any intention to cross the continent to the wilder Pacific shore. New York was much more like a European city than anything he could meet in the West—urbane, sophisticated, and filled with an extra life that seems to have spurred Muybridge to cross the Atlantic. Although the West was the place for risk takers, those who took a chance with their lives in a literal pursuit of gold, New York provided a good compromise between excitement and civilization.

It is not clear how, but Muybridge managed to get the job of commission merchant for the London Printing and Publishing Company. As such he was not only the publisher's agent but also a subcontracting publisher in his own right. He took unbound books—sheaves of printed paper—from LPPC and had them bound, then sold them within the United States. This was a complex role for a man in his early twenties with no experience of publishing. Not only had he to arrange the shipping and production, he also had sole responsibility for marketing the books. If Muybridge did not make a good enough sales pitch he would operate at a loss. He seems to have made a success of it, though. In a statement at Muybridge's murder trial, a New York friend, Silas T. Selleck, would comment that Muybridge was a good businessman. (Of course, the accuracy of any character statement made at such a trial could be questioned.)

Selleck may have been more than just a friend to Muybridge; perhaps an inspiration. Selleck was an early photographer, using the daguerrotype method, soon to be outmoded by the more flexible wet collodion technique. It is possible that Muybridge experimented with photography while in New York—certainly Selleck was the first photographer with whom he had significant contact—but at the time, Muybridge seems not to have considered it a career.

His friend Selleck, caught up with the gold rush fever, moved to San Francisco, and set up a studio there to keep up an income while he tried his luck at prospecting in between sittings. Muybridge must have found the opportunities to take similar advantage of the new markets of the West tempting. Why shouldn't he ply his trade in books in California while seeking the instant fortune of the lucky gold hunter? Around the end of 1855, the temptation became too much. Muybridge took the trail out West.

Muybridge leaves no account of the sheer effort involved in crossing from New York to California. It may have been that he took the safer if lengthy sea route, but we do know that he was to return overland from San Francisco in 1860. This was no simple journey. Not only were there many changes of coach and horses along the way, it involved passing through hundreds of miles of backwoods country with no towns, no civilized facilities. Practically every journey would involve overcoming major difficulties, and it was a common enough occurrence for coaches to become stuck or to overturn.

Luckily, we do have a very detailed account of a similar journey in Horace Greeley's *An Overland Journey from New York to San Francisco in the Summer of 1859.* Greeley was a famous character in his time. Founder of the *New York Tribune* newspaper, he was a great believer in the emancipation of slaves and of the opening up of the country; it was he who popularized (though did not, as is often thought, originate) the phrase, "Go west, young man."

Although Greeley's journey involved a few detours that the dedicated traveler to California would not have undertaken, it still shows both the effort required of the traveler and the rough conditions along the way. At one point, Greeley bemoaned his gradual loss of luxury.

> I believe I have now descended the ladder of artificial life nearly to its lowest round . . . the progress I have made during the last fortnight toward the primitive simplicity of human existence may be roughly noted thus:
>
> May 12th—Chicago—Chocolate and morning newspapers last seen on the breakfast-table.

23d—Leavenworth—Room-bells and baths make their final appearance.
24th—Topeka—Beef-steak and wash-bowls (other than tin) last visible. Barber ditto.
26th—Manhattan—Potatoes and eggs last recognized amongst the blessings that "brighten as they take their flight." Chairs ditto.
27th—Junction City—Last visitation of a boot-black, with dissolving views of a board bedroom. Beds bid us good-by.
28th—Pipe Creek—Benches for seats at meals have disappeared, giving place to bags and boxes. We (two passengers of a scribbling turn) write our letters in the express-wagon that has borne us by day, and must supply us lodgings for the night. Thunder and lightning from both south and west give strong promise of a shower before morning. Dubious looks at several holes in the canvas covering the wagon. Our trust, under Providence, is in buoyant hearts and an India-rubber blanket.

All rather different from the neat, suburban orderliness of Kingston-upon-Thames. Alongside the decline in luxuries was an increasingly uncomfortable mode of transport. Through to Kansas, taking a mere six days from New York, travel was on the newly opened railroads, with only a trip down the Mississippi by steamer from Quincy to Hannibal to break the rail link. After that it was a very different venture. The stages and wagons that threaded the rough roads were much slower and much less reliable. Greeley described an all too typical experience of crossing a river with no bridge available:

Coming to Turkey Creek, the passengers were turned out (as once or twice before) to lighten the coach, which was then driven through the steep-banked ford, while the passengers severally let themselves down a perpendicular bank by clinging to a tree, and crossed a deep and whirling place above the ford, on the vilest log I ever attempted to walk—twisty, sharp-backed and every way detestable. One of the passengers refused to risk his life on it, but hired one of the lazy Indians loafing on the further bank to bring over a pony, and let him ride across the ford.

Greeley, the city man, was not impressed by the scale of townships in Kansas. After experiencing Prairie City, Peoria City, and Ohio City, he commented, "It takes three log houses to make a city in Kansas, but they begin *calling* it a city as soon as they have staked out the lots." He finally arrived in San Francisco just under four months after leaving New York—and though all his detours weren't necessary, the fact remains that Muybridge would have taken months rather than weeks to cross to the West Coast.

The length and difficulty of crossing is something that Greeley emphasized. When speaking of California, and San Francisco in particular, he noted that the "resource" that was most lacking was women, and that there was a great need for a railroad to link San Francisco to the East Coast (he gives us a whole chapter on how the financing of this could work). As he said of San Francisco:

> A railroad communication with the Atlantic states would make her the New York of this mighty ocean—the focus of the trade of all America west of the Andes and the Rocky Mountains, and of Polynesia as well, with an active and increasing Australian commerce. Without an inter-oceanic railroad, she must grow slowly, because the elements of her trade have been measured and their limits nearly reached.

These limits, he believed, were largely imposed by isolation. Greeley also described what the San Francisco that Muybridge first saw was like.

> I estimate the present population at about 80,000. . . . San Francisco has some fine buildings, but is not a well-built city—as, indeed, how could she be? She is hardly yet ten years old, has been three or four times in good part laid in ashes, and is the work mainly of men of moderate means, who have paid higher for the labor they required than was ever paid elsewhere for putting so much wood, stone, brick and mortar into habitations or stores.

One of the stores, 113 Montgomery Street, belonged to Muybridge, or rather Eadweard Muygridge as he first styled himself in mid-May of 1856. It is possible that this strange variant resulted from an accidental misspelling of Muggeridge, but more likely it was a conscious modification given the way Muybridge had already changed his Christian name, perhaps to give his name a form more easily dealt with in the Californian world of crossover between Spanish- and English-speaking cultures.

Muybridge continued to act as an agent for the London Printing and Publishing Company and also took on the East Coast firm of Johnson, Fry and Company. His work for publishers took place in an upstairs room, while at street level he operated a bookshop. He was clearly having some success, as an advertisement in the April 28, 1856, edition of the *San Francisco Daily Evening Bulletin* tells us that he was looking for "a gentleman well qualified to obtain subscribers for a new illustrated standard work. None but a first rate canvasser need apply."

He had tired already of the legwork involved in being a publisher's sales agent and was doing well enough to take on help to sell a new illustrated edition of Shakespeare from the Martin Johnson Company of Philadelphia. Interestingly, this advertisement provides a missing link between his names of Muggeridge and Muygridge—here he calls himself "E. J. Muggridge." In a later flyer dating from around 1859, he also offered a service to "gentlemen furnishing libraries" to find and purchase books for them sourced from Europe and the East Coast.

It might seem rather incongruous that Muybridge could make a living from such a refined business in what we might think of as a wild frontier state, but California had an unusually rich mix of population from gold diggers to entrepreneurial millionaires. There was a huge interest in reading; Greeley observed that there were between ninety and a hundred periodicals published in California, around a third of these issued in San Francisco, and this in a state that was less than 10 years old.

Muybridge was also not above making money out of the less salubrious aspects of San Francisco life at that time. In August of 1856, now as Muygridge, he placed this advertisement:

SEVERAL ATTEMPTS HAVE RECENTLY BEEN made to furnish the friends of JAMES KING OF WILLIAM, and the public, with a faithful and striking likeness of him; but almost total failure has resulted. The portraits which have been published are insignificant works of art.

But Muybridge, it seems, was going to issue on September 1st a "large and beautifully executed portrait of the late Mr. King." printed from the stone at $2.50 on India proof paper or $2 on plain white paper. Once again he was on the lookout for agents to sell his product. James King was not an obvious subject for a popular portrait. He was neither a celebrity nor a statesman, but a newspaper editor. King had taken a tough stance on lawlessness and had been assassinated back in May for his pains, which transformed him into the sort of hero whose portrait would sell well.

Muybridge's success on America's West Coast was to be felt all the way back in Kingston—but not in an entirely positive way. For Muybridge's mother, Susannah, her oldest surviving son's good fortune was to result in her desertion by the rest of her children. John was already dead, and Edward's two younger brothers, George and Thomas, both followed the enterprising Muybridge across the continent to San Francisco. Probably their intention had originally been for George to take charge of the business while Muybridge himself had more of a roving brief, but George already suffered from consumption (tuberculosis) when he traveled out to California and would be dead by 1858. Tom had been working as a seaman, but seems to have been called over to replace George, so by 1859 two of Susannah Muggeridge's four sons were dead, and the other two were living in one of the remotest parts of the world.

Once Tom was established in the store, Muybridge was free to make more excursions around California, where he became fascinated with the natural magnificence of the newly discovered Yosemite. His enthusiasm for photography was waxing, and he saw an opportunity to make money out of this incredible natural resource, capturing lifelike images with the beguiling technology of the photograph.

∞∞

With our visual sense now flooded by motion pictures, TV, and glossy color images, it's difficult to remember just how stunning a step forward the introduction of photography was. As a blend of art, science, and technology it was a particular temptation for the creative mind of Eadweard Muybridge.

In 1860, when Muybridge first considered a career in photography, the whole business was little more than 20 years old, although the two key components that were fused together in the photographic process were known long before; it was the practical combination of the two that it took nineteenth-century technical ingenuity to realize. To make a photograph it was necessary both to get an image of the real world projected onto a surface and to cover that surface in a light-sensitive material to pick up and reproduce that image.

The solution to the first part of the challenge, projecting an image of the world, was found in the ancient technology of the camera obscura. The most basic camera obscura—a darkened room with a pinhole in the shutter or wall to let in a tight beam of light—was known 2,500 years ago by the Chinese philosopher Mo Ti. It was Mo who first realized that light moved in straight lines. This meant that light from the top of an object would pass through a pinhole and form the bottom of the image, while the opposite happened to light from the lower parts of the object, resulting in an upside-down picture displayed on a surface placed in the path of the incoming light. Mo didn't invent the device, but his was the first known explanation of how it worked. Mo Ti lived in the fifth century BC. In the West around 330 BC, the Greek philosopher Aristotle also observed this effect and noted how the sharpness of the image increased as the size of the hole got smaller.

In medieval times the camera obscura was improved by replacing the pinhole with a glass lens, and from this time on to the Victorian era the only real enhancements made to the technology were more complex optics, which enabled focusing without moving the screen the image was projected on, and employing mirrors so the image could be displayed on a horizontal table. These early devices were effectively part of a room, though more portable devices relying on a tent or a box to provide a dark environment (the image is very faint) were increasingly used beginning in the 1500s.

The development of portable camera obscuras (at the time simply known as cameras) was driven by the increasing variety of applications that were found for them. Astronomers realized that they provided a safe way to observe the Sun without staring directly at the blinding rays. Artists, from the seventeenth century onward, also found them a handy aid, projecting the image to be painted or drawn onto a surface, where it provided a clear guide for the pencil or brush. (Vermeer, for instance, is thought to have made significant use of the camera obscura in his paintings.)

It was this use of the camera obscura, to help the artist, that led William Henry Fox Talbot to invent the familiar negative-based photographic process—but we are getting ahead of ourselves. It is one thing to project an image; it's another to freeze that ever-moving image in time, capturing it automatically onto a surface. What was needed was a substance that changed when it was exposed to light, a way of preserving the image after the camera obscura stopped projecting it.

The property of sunlight to make changes in physical materials— to burn the skin, to change the color of wet soil as it dries—was a known fact for thousands of years, but the earliest record we have of a specific change being triggered by light that was both quick and visible was made by the Roman engineer Vitruvius, who noted that red lead, one of the oxides of that heavy metal, turned black in sunlight. For photographers, though, the essential photochemicals for capturing a visual image would be the compounds of silver.

The earliest reference to the light sensitivity of a silver salt (probably silver chloride, the compound of silver and the poisonous gas chlorine) appears in the writing of Pliny the Elder, a Roman scholar who lived from AD 23 to AD 79. Later, during the eighth century, the Arab Abu Musa Jabir Ibn Hayyan, known in the West as Geber, was to observe a similar darkening in silver nitrate. Although a number of workers on the cusp of the change between the unstructured wildness of alchemy and the science of chemistry, notably Georg Fabricus, Angelo Sala, and Robert Boyle, all made similar observations, it wasn't until Johann Heinrich Schulze got to work on this odd behavior around 1725 that it was definitively pinned down to being a reaction of the substance to light.

Schulze was to inspire the development of photography's immediate predecessors, as he went beyond observing a change in color to make practical use of this phenomenon (if in a decidedly dilettante fashion). Schulze cut templates out of card, which he wrapped around bottles containing silver nitrate, forming darkened shapes in the chemical, and wrote with silver nitrate solution on paper, producing a secret message that became visible only when exposed to light.

The trouble with Schulze's light pictures in a bottle was that they could not be kept for very long. The light that made the picture visible also darkened the surrounding silver nitrate until the bottle was filled with a dull, blackened mass of crystals. Over the next hundred years, small jigsaw pieces of knowledge would be fitted into the picture that would eventually become working photography, but it was only very close to the end that the true potential was spotted.

In the 1770s, Swedish chemist Carl Scheele was the first to realize that the blackening of a silver salt in light (he was working with silver chloride) was produced by the formation of tiny particles of metallic silver. In fact, he also found out that he could stop the process from running away and destroying any image produced by washing a surface coated in silver chloride with ammonia. This technique removed the salt, but left the blackened silver behind. No one seemed to notice the significance of this discovery.

Scheele's experiments were taken further in 1802 by the formidable British duo of pottery magnate Thomas Wedgwood and scientist Sir Humphrey Davy, who had already been working on a method of producing sun prints. These images (now called photograms) are a more sophisticated version of Schulze's light bottles. A solution of silver nitrate was spread onto a surface like paper (though they found that the best results came from a base of pale leather), then exposed to light, but with objects from ferns to keys placed on top of the material to cast a shadow and hence produce an image.

Wedgwood, who was the driving force behind the experiments, was an enthusiastic amateur scientist—this was a time when the rich dilettante could still make a genuine contribution to scientific knowledge—though it's possible he also had a practical application in mind, hoping to develop a process for making images on his pottery. But Wedgwood and Davy never worked out a way of protecting the image, fixing it so that it wouldn't disappear in sunlight. Wedgwood even experimented with what would have taken practically the last step to the photograph, projecting the picture from a camera obscura onto some light-sensitive material, but concluded it wouldn't work because the image was too faint.

Wedgwood was the son of a famous father—Josiah, the founder of the pottery—but it was John Herschel, son of the better-known William, who found the solution to Wedgwood and Davy's problem. Science was part of everyday life for Herschel, who was brought up in a laboratory. His father's home, Observatory House, built by William near Windsor in England after he had discovered the planet Uranus, was also a workplace. John could hardly avoid the scientific world. He became a fine astronomer in his own right, but like his father dabbled in all the sciences.

John Herschel used sodium thiosulfate (a less unpleasant chemical than ammonia, known in the photographic trade as hypo—the old name for the compound was hyposulfite of soda) to wash out the unchanged silver salts, which, he commented, dissolved as well in hypo as sugar dissolves in water. Herschel would also coin some of the familiar photographic jargon—"photograph" itself, and "negative" for the brightness-reversed image produced in the camera—but that would come later.

As the nineteenth century got under way it took four last pieces in the puzzle to reach the photographic technology used by Muybridge, pieces held by Niépce, Daguerre, Fox Talbot, and Archer.

French inventor Joseph Nicephore Niépce was another, even richer, dabbler. He was originally looking for a more effective way to produce an image on a printing plate, as opposed to engraving the picture to be printed by hand. Niépce tried out silver salts like everyone else, but didn't achieve any more than his predecessors. Frustrated, he turned to an unlikely substance called bitumen of Judea. This heavy, oil-based material has two interesting properties. Black in its natural state, it becomes lighter when exposed to sunlight. The exposed portions set into a hard form, while the unexposed material remains fluid and is easy to remove with a suitable solvent. By 1822, Niépce was producing crude images this way, using exposures of around eight hours, and in 1826 or 1827 captured what is generally regarded as the first direct image (rather than true negative-based) photograph, or heliograph as he called it, from the window of his house.

Niépce did not take his invention any further, and the bitumen-based image was something of a dead-end technology. It required immensely long exposures and could be viewed only under very careful conditions—the image was visible only when light struck the picture at a certain angle. Even so, this was the first photographic image, and Niépce deserves more recognition than he usually gets outside France. But a friend of Niépce's, the artist Louis Jacques Mandé Daguerre, saw a great potential for the technique if it could be made easier to use.

Daguerre learned a lot conceptually from Niépce, though in practice he went back to using a silver salt. In his process a copper plate was coated with silver, then "sensitized" by placing it in a box of iodine vapor, which left a thin layer of silver iodide on the surface of the plate. This silver salt was exposed to light, then made to change color (develop) using mercury vapor, an insidiously dangerous substance.

Finally, the resultant image was fixed—removing any excess silver iodide—with a hypo solution. Just like Niépce's heliographs, the result was a direct positive picture, but with an exposure that could be as little as half an hour (over the next 20 years, by using better lenses, this time would be brought down to about a minute). Daguerre modestly called his new images daguerreotypes.

The penultimate piece in the photographic puzzle came from another dilettante scientist, William Henry Fox Talbot. Fox Talbot had been brought up in the delightful rural English setting of Lacock Abbey. Built as an Augustinian nunnery on Wiltshire meadows running down to the river Avon, the fifteenth-century buildings became a private house after Henry VIII closed down the monasteries. This rambling, fascinating home had been Fox Talbot's playground when he was a child and would become his inspiration.

Fox Talbot was certainly influenced by others who had worked in the field, but he did not know about the daguerreotype process (he published a paper on his "photogenic drawing" process, producing a "calotype" eight months before Daguerre went public). Fox Talbot also used silver salts, but unlike Niépce and Daguerre he did not go out of his way to find a method of producing a "true" positive image, hence losing the need for the dangerous mercury vapor stage in the daguerreotype. Instead, he turned a problem into an opportunity. He intentionally produced a negative image, on translucent paper or a glass plate. Light was then shone through this negative to expose a second photosensitive material—treated paper, say—to produce a positive image. This meant that you could produce as many copies as you liked. In 1835, Fox Talbot successfully developed his first calotype, like Niépce's first picture a scene from the window of his home, in this case the view from the elegant oriel window of Lacock's south gallery.

The final, and most obscure, name of the founding fathers of photographic technology was that of Frederick Scott Archer. Unlike Fox Talbot, Archer started life as a poor orphan, originally apprenticed to a silversmith in London. He became interested in coins, and then sculpture, his work finding favor with the fashionable set. Archer used photography as an aid to his sculpture, but found the processes of the day unsatisfactory. In 1849, Archer devised the wet collodion process that would become the standard photographic technology of Muybridge's early years behind the camera. Unlike most of his competitors, he never patented the process, so he gained no rewards from the technology that wiped out daguerreotypes and calotypes, and he died penniless in 1857. We'll come back to how the process worked, but it had the flexibility of the calotype, yet produced a much clearer image, and would work with an exposure of just a few seconds, significantly faster than a calotype.

Apart from Niépce's side step into using bitumen, all the ancestors of photography and the real processes Muybridge would use depended on the way that silver compounds darken when exposed to light. To get good, detailed results, the images need to be produced using very small particles of the selected silver salt (typically silver chloride or silver bromide)—tiny crystals or grains. The finer the grains, the more detail in the photograph but also the less sensitivity to light. To understand what's happening in the silver salt it is necessary to get down to the submicroscopic level of molecules and of photons of light, so it is hardly surprising that good explanations of the science of photography lagged well behind the practical use of the technology.

The initiator of the photographic process is a photon, a packet of electromagnetic energy that crosses the intervening space from the subject of the photograph to the light-sensitive material and hits a silver salt crystal. As often happens, the photon interacts with one of the electrons in the crystal. In effect, the photon is destroyed and its energy is absorbed by an electron within the crystal of silver salt. Electrons can only operate at specific energy levels, rather than in a continuum of possibilities, so the photon must have sufficient energy to force the electron to jump to a higher level—a quantum leap to an excited state. If this were a simple molecule of silver chloride or bromide, very soon the electron would drop back down, re-emitting a photon of light. But the crystal has a more complex structure, including some locations—faults, effectively—where the excited electron can get trapped. In the trap, the electron has a chance to react with one of the positive silver ions in the crystal, turning it into a metallic silver atom.

Of itself, the single silver atom won't have any visible impact. It's too small. But after the image has been exposed, some of the crystals in the image will have one or more silver atoms in them, and some won't. When the image is then developed, the developing solution also converts silver salts to silver, but only in crystals, where this process is catalyzed by existing silver atoms. The more metallic silver there is in a crystal, the quicker it darkens. The exposed silver specks pull the whole grain along with them to darken during the development process. Finally, the picture is fixed by removing any excess silver salts that have not reacted—often still using hypo.

∞∞

By 1860, though no one knew how they worked, photographs had become an important part of everyday life. By now Eadweard Muybridge was ready to leave his agency business fully in the hands of his brother Tom, and to sell or rent out the shop, freeing himself to travel and to put his photographic plans in action. In an advertisement in the *San Francisco Daily Evening Bulletin* for May 15, 1860, he made his intentions public:

> I have this day sold to my brother, Thomas S. Muygridge, my entire stock of Books, Engravings, etc. and respectfully request a continuance of public patronage in his favour.

He wanted to feel out the potential audience for photographs of the dramatic American landscape and also to undertake a buying trip back in Europe, as he went on to say:

> After my return from the Yosemite I shall on 5th June, leave for New York, London, Paris, Rome, Berlin, Vienna, etc., and all orders or commissions for the purchase of Works of Literature or Art entrusted to me, will be properly attended to on the following terms: All amounts under $500 10 per cent; between $500 and $1,000, 5 per cent; over $1,000 2 $\frac{1}{2}$ per cent.

He had intended to take the steamer *Golden Age*, which left San Francisco on June 5th, but he was held up, perhaps overwhelmed by his visit to Yosemite. Instead he took the overland route back toward New York, something we can be certain of, because the result was a catastrophe. As Horace Greeley found in his journey, traveling this way gave a good opportunity to really appreciate the wild landscape, and perhaps that was why Muybridge decided to rough it. He was to see more than he bargained for. The ill-made trail through Texas to Ar-

kansas was to prove too much for the Butterfield mail coach in which
Muybridge rode.

It had been obvious ever since the coach left Mountain Station in
the charge of a driver and one Mr. Stout, a roadmaster of the Butterfield
line who was acting as conductor, that the new team of horses was
skittish. It's not really surprising. According to Muybridge these were
"six wild mustang horses"—hardly a calm, controlled team. From the
first crack of the driver's whip they broke into a run. About half an
hour later, as the coach reached the highest point of the trail over the
mountain, the driver pulled on the brake lever to hold the coach back
on the steep decline ahead. Nothing happened. The crude wooden
brakes might as well have not been there.

It was only a matter of time before disaster struck. Muybridge
showed surprising coolness of thought. To attempt to jump from the
side of the coach was risking death—in fact another passenger (named
Mackey, May, or McCarty in different newspaper reports) was killed
this way when he attempted to leap to safety. Instead, Muybridge took
his knife and began to hack at the canvas back of the coach to try to
ease his way safely to the ground behind the careering vehicle.

The headlong charge continued. The driver managed to maneuver
the horses off the road, thinking that the rougher terrain would slow
them down. A wheel struck a rock, smashing the wooden spokes to
matchwood. The coach swerved violently, tipped, slammed into a tree,
and shivered into pieces. Muybridge was thrown free of the wreckage.
His head hit a rock and he lost consciousness.

When Muybridge next became aware of his surroundings, he
found himself in a hospital bed. He could not remember lying unat-
tended in the rough scrub for hours before a rescue wagon that was
searching for the missing mail coach picked him up. His senses were
shattered. His head was wracked with pain, his skull fractured. It was
only several days later that he discovered he had been taken to the mili-
tary station at Fort Smith, 150 miles from the scene of the accident.
Fifteen years later, Muybridge recalled coming round:

> There was a small wound on the top of my head. When I recovered each
> eye formed an individual impression, so that, looking at you, for instance, I
> could see another man sitting by your side. I had no taste nor smell, and

was very deaf. These symptoms continued in an acute form for probably three months. I was under medical treatment for over a year.

Muybridge's recovery might have been gradual, but within days of regaining consciousness he declared himself well enough to travel on as far as New York. There he attempted to get further treatment while taking legal proceedings against the Butterfield Overland Mail Company. But Muybridge was not happy with the medical care he was receiving and traveled still further, back to London. He was still suffering from double vision and impaired sensory responses.

His London doctor, the acclaimed Sir William Gull (at the time Queen Victoria's physician), was well known for his enthusiasm for outdoor activity as a cure-all and is likely to have prescribed a long break with a gradual increase in exercise and healthy fresh air. He also would have suggested giving up meat, which was generally regarded as an impediment to recovery. Muybridge had no problem with the encouragement to take exercise. Although it would take him several years to fully recover, he could not stay inactive long.

We know that early in 1861 he returned to New York to continue his legal battle with the coach company—in the end he settled for $2,500 (the equivalent of around $50,000 now) though he had initially asked for four times as much. But for the next five years he was mostly based in England, even though he continued to travel widely. In part, this may have been because Muybridge felt more comfortable on home territory while recuperating, but it is most likely, given the duration of the American Civil War (from 1861 to 1865), that Muybridge was prudently sitting out the devastation in his adopted country.

During his time in England, Muybridge's interest in photography continued to grow, while his natural ingenuity blossomed into practical creativity with the production of a series of inventions. In the early 1860s, he registered two patents that provide a fascinating peek into this largely undocumented portion of Muybridge's life.

The first of these documents presents us with a mystery. It is a patent for a printing process: "an improved method of and apparatus for plate printing" by E. Muggeridge of New York. It is dated September 28, 1860, and was sealed on March 8, 1861. There are a number of oddities here. Muybridge had started calling himself Muygridge rather than Muggeridge several years before 1860. Technically he may have been in New York in September 1860; we aren't sure exactly when he went to England. He would still have been in a poor physical condition and was hardly likely to be dashing off patent applications. What's more, he was merely a transient in New York; he would properly be "of San Francisco."

There seem to be two possibilities. One is that Muybridge devised this technology while living in New York back in the mid-1850s, and it had taken a long time for it to get through the patent authorities' review process. The other is that his cousin, Edward Smith of New York, who had a small printing press in his stationery shop, may have decided to use the Muggeridge name rather than his own. Of the two possibilities, the first seems more likely, as there is one piece of evidence that suggests that the invention, which overcame the problem of having to wipe printing plates clean of ink before making each impression, really was Muybridge's idea. In 1861, he wrote to his uncle Henry Selfe, who had emigrated from England to Australia. He offered Selfe the opportunity to freely exploit two inventions, provided Selfe paid to have them patented in Australia. Muybridge did make one proviso, though: "Should, however, you realize 'something handsome' from the undertaking (of which I am almost fully persuaded you can), or should I at any time be in need of money, some mutual understanding may then perhaps be arrived at."

The first of the inventions Muybridge described to his uncle was a plate printing apparatus. This corroborates Muybridge's authorship of the Muggeridge plate printing process patent, even though it leaves the wording a mystery. The second invention he asked Selfe to consider (the subject of Muybridge's other patent) demonstrates how his mind darted from subject to subject. Rather than involving the book trade or a contribution to his new interest of photography, it is for a washing machine. The patent is dated August 1, 1861, around the time of Muybridge's return from New York. Perhaps he devised it on the boat on the way back across the Atlantic. He wrote his letter to Henry Selfe two weeks later, though the patent would not receive its seal until January 24, 1862.

There is no evidence that Selfe ever took up Muybridge's offer, and his uncle was to die six years later, aged 62. According to his gravestone, he "drowned during a flood in the Warragamba."

The washing machine has a much more logical attribution. The patent is ascribed to "Edward James Muygridge of San Francisco, at present residing at No. 6, St. John's Villas, Adelaide Road, St. John's Wood, in the County of Middlesex." The form of name is right for the period, as is the address. This, the patent proudly claims, is a "peculiar construction and arrangement of machinery or apparatus, whereby the operation of washing clothes and other articles of a textile nature is greatly facilitated."

In design, Muybridge's machine looks decidedly evil. The action takes place in a curved wooden trough, like a section out of a barrel. A hand-turned wheel with a crank action moves two large "pounders" back and forth, up and down the curve of the trough, squashing the items of clothing against a ridged section that Muybridge describes as being "corrugated, similar to an ordinary washing board." It's like two huge meat tenderizer mallets side by side, squashing the clothes into the corrugated boards.

Muybridge was staying with his aunt, Mrs. Rachel Wyburn, one of his mother's sisters. His mother had also been staying there, but returned to Kingston around this time to live with John Smith, her brother, who had never married. It's not known whether Muybridge chose to move to independent accommodation or to spend all his time in England with his relatives.

It has been suggested that Muybridge's growing interest in photography, particularly landscape photography, was a result of his doctor's recommendation of outdoor activity as a means of recuperation, but it should be remembered that before his accident, Muybridge had already decided to branch out into this new business, and that his disastrous return journey to Europe had been in part to check out the market for photographs of the American wilderness. During the few years Muybridge was recovering, photographs had been transformed from a novelty into a household necessity. In particular, an invention called the "stereoscope" had made photographs the Victorian equivalent of today's TV—an essential part of life and entertainment in the nineteenth-century parlor. The time seemed right to make his interest into a business.

The stereoscope is a rare example of a technology that flourished over 100 years ago yet still seems remarkable today because it is no longer commonplace and because its successors have never had the same commercial success. Instead of one photograph, a stereoscopic camera takes two simultaneously, with a separation that parallels the separation of the eyes, so each picture presents a subtly different view. We see a three-dimensional view of the world because the brain combines "flat" images from the two eyes, separated by around $2\,^1/_2$ inches, and uses their slightly altered viewpoint to detect depth. The two simultaneously taken images in a stereograph are viewed through the stereoscope, which presents one picture to each eye, allowing the brain to combine them in the same way.

At one time it was thought that the optic nerves that link each eyeball to the brain were connected purely into opposing sides of the organ (so the right-hand side of the brain received signals from the left eyeball, and vice versa), but this would make it very difficult for the brain to make the high-speed comparisons necessary for stereoscopic binocular vision. It was discovered in the 1990s that in a fetus, while all the neurons growing from the eye begin by crossing toward the opposite side of the brain, some are repelled by a special protein and end up connecting to the same side of the brain as the eye.

The more growing neurons that bounce back from the midline of the brain, the better the binocular vision. Around 40 percent of our connections don't cross over, as opposed to 3 percent in mice, and none at all in many birds and fish that totally lack binocular vision. Because each side of the brain, handling its opposing eye, has information from the other eye to compare, between them they can build up a comprehensive picture that gives us a convincing illusion of seeing in three dimensions. Each half of the brain still handles information primarily from the opposite side of the field of view, but this information comes from both eyes.

The idea of presenting a different image to each eye to get a more realistic picture predates photography. Back in the late sixteenth century, Giovanni Battista della Porta and Jacopo Chimenti da Empoli produced side-by-side drawings, and in 1613 the French clergyman Francois d'Aquillon came up with the term "stereoscopic" (or to be more precise, *stéréoscopique*); but without an optical device to control the view of each eye it is very difficult to separately focus the two eyeballs, and the limits of accuracy of drawing by hand made the whole thing impractical until photography was invented. However, it was just a matter of years after the earliest true photographic image was made before a working stereoscope was built.

The device that was to become so popular in Victorian households was invented in 1838 by Charles Wheatstone, the British challenger to Morse in the attempt to build the first electric telegraph. Wheatstone gave it both of the names that would attach to it over the years—the stereoscope and the stereopticon—though he never went beyond a theoretical description of the process. It wasn't until 1849 that David Brewster made use of the fledgling photographic technology to build the first stereoscopic camera. It featured twin lenses, separated just like a pair of eyes, to capture a pair of images on the photographic plate behind it.

Just as a celebrity endorsement might raise awareness of an entertainment product today, it was Queen Victoria's enjoyment of the stereoscope at the Great Exhibition, held in the purpose-built Crystal Palace in London in 1851, that turned the stereoscope from a novelty to a huge commercial success. Within five years, over a million homes in the United Kingdom owned a stereoscope, and this enthusiasm for three-dimensional pictures was soon reflected across the Atlantic in the United States, where the technology was promoted by Oliver Wendell Holmes.

Stereoscopes went on to have a long and checkered history. In later years they became widely employed in aerial photography, which since the mid-1940s has depended on the extra information provided in a stereoscopic image to make it easier to identify features on the ground. The stereoscope as entertainment saw its final outing to date in the View-Master system—viewers that hold a circular cardboard disc with tiny pairs of stereoscopic images. The viewer changes the image on display by pressing a lever to rotate the disc. These toys were hugely popular in the second half of the twentieth century, and though less common now, are still available from the retail giant Mattel. Stereo-scopic images have not gone away, though we're more likely to see 3-D movies these days, but the newer three-dimensional processes have never recaptured the excitement that the stereoscope provided in the Victorian drawing room.

Although there is no evidence that Muybridge was directly inspired by the stereoscope, he used stereoscopic cameras in much of his work, especially in the motion studies that made him famous. Such was the popularity of the stereoscope that if you wanted to sell photographs to a wide audience it was taken for granted that they would be stereo-scopic images.

Around 1866, with the aftershocks of the Civil War fading, Muybridge made the long journey back to California. The chances are that, rather than risking another overland trip, he returned by the in-creasingly popular mode of travel of steaming down the East Coast, crossing the Isthmus of Panama by train (the canal was not yet built), and taking a boat up the West Coast to San Francisco. With him he took the latest photographic equipment, bought with the profits of his book business and the outcome of his lawsuit. He also took one further thing: another new name. It may have been that Muygridge had worked well in California but proved too much of a tongue twister in the United Kingdom, and thinking of future sales he decided on a more conventional final syllable. From now on he would be Eadweard Muybridge.

THREE

ASSEMBLING IMAGES

The beauties and wonders described in this book, however, are not presented for the benefit of the sick, but to the crowd of pleasure-seekers who make their annual visitations to Niagara, Newport, Saratoga, Cape May, and other centers of fashion, frivolity, foppery and folly. With half the expenditure of money and vital force thus thrown away, to the moral and physical deterioration of all concerned, the California trip, via the Pacific Railroad, may be thoroughly enjoyed. There is nothing in it to enfeeble, but everything to strengthen.

From Samuel Kneeland's *The Wonders of the Yosemite Valley and of California* (1870)

The San Francisco to which Muybridge returned at the age of 36 had moved on in the handful of years he had been away—hardly surprising considering the extreme youth of the city. It was more sophisticated and better established; its population had doubled. Many of the businesses of Muybridge's time had folded or changed premises. The remains of his own establishment, the book agency run by his brother, had shut down when Thomas opted for further emigration, out to Australia, perhaps at the recommendation of their uncle. There Tom seems to have converted to dentistry, by the simple act of buying an old surgery and putting up a shingle. It doesn't make an encouraging picture for his clients. The skills overlap between book agent and dentist is limited. But Tom persevered. Like his brothers, though, he seems to have tired of the name Muggeridge: He changed it to the simpler, slurred "Muridge."

Muybridge, meanwhile, with no commercial base in the new San

Francisco, where it was now less easy to simply slip into premises, spent a year squatting in the gallery of his old friend Silas Selleck, who had abandoned the obsolete daguerreotype technology in favor of the more modern technique of wet plate photography. Where Selleck concentrated on the immediately lucrative business of portraits, however, Muybridge was still determined to capture the landscape and the special character of San Francisco itself.

This was not a period when you could slip your camera from your pocket and take an instant photograph. Not only was the camera itself bulky and in need of solid placement on a heavy-duty tripod, but the technical aspects of the process were also messy and time consuming. Although Frederick Scott Archer's wet collodion process was much simpler (and less dangerous to the health of the operator) than the daguerreotype that it made obsolete, it still took considerable time and expertise on the part of the photographer. In fact, the process was so complex, it's surprising that it was used so successfully.

The starting point was a plate of glass—ordinary window glass, but cut to the required size, scrupulously cleaned by immersing it for an hour or more in nitric acid or a strong alkali, left to dry, polished with a suspension of very fine pumice in alcohol, and dried with a chamois leather. Before the active chemicals were added to the surface, it was often treated, either by coating the edge with a rubber solution or giving the plate a thin coating of albumen (egg white) to help the chemical mixture stick to the surface.

The glass then had to be coated evenly with a chemical mixture in a dark place. Within minutes, before the mixture could dry, the plate had to be placed in the camera, exposed, and then immediately processed to avoid losing the image. It could take as much as an hour to successfully complete the production of a single negative. Muybridge described this process himself in his book *Animal Locomotion*, written after pre-prepared dry plates had become common.

> Every photographer was, in a great measure, his own chemist; he prepared his own dipping baths, made his own collodion, coated and developed his own plates, and frequently manufactured the chemicals necessary for his work. All this involved a vast amount of tedious and careful manipulation from which the current generation is, happily, relieved.

Not only was this process tedious, it was still potentially fatal. Admittedly the photographer was not exposed to the deadly mercury vapors of the daguerreotype process, but the collodion Muybridge refers to was gun cotton, the result of soaking finely carded cotton in a mixture of nitric and sulfuric acid. The modified cotton, now highly explosive, had to be washed, dried, and then dissolved in a mixture of ether and alcohol to produce a dangerously flammable gel. This gel was then iodized by adding in iodine salts, or more often a mixture of iodides and bromides.

The iodized collodion could be stored for several months (and had to be left to stand for a day or two before use). When the photographer was ready, the glass plate was held by one corner and the gel poured on from the opposite corner. Part of the skill of making wet plates was getting the collodion to settle evenly across the glass, requiring a careful, rhythmic rocking to disperse the gel as consistently as possible.

The resultant gooey sheet of glass was sensitized by sliding it into a bath of silver nitrate solution, which reacted with the iodine and bromine salts to leave light-sensitive silver iodide and bromide crystals suspended in the collodion gel (from this stage onward, the plate would need to be kept in darkness). The collodion acted as a matrix in which the silver salts gradually became suspended, a process that took several minutes. This suspension would later be termed an emulsion.

While the plate was still slightly damp—after sensitizing, the plate had to be used within around 10 minutes—it was put into the camera, still in darkness, and exposed. Finally, the plate would be developed in a bath, based on ferrous sulfate, that darkened the most exposed areas, then fixed in another bath of hypo or potassium cyanide (which worked better but was highly dangerous) to remove the excess silver. Rinsing, drying, and varnishing for protection finished off the lengthy procedure.

So complex was the process that the author of *Alice in Wonderland*, Lewis Carroll, an accomplished early photographer, thought it wise to add some extra verses to a poem he had written on photography in the style of Longfellow's *The Song of Hiawatha*. Carroll's original poem described the process of taking the photograph, but he added these

lines when he thought that the complexity of the wet collodion process
might be forgotten by later generations:

> First, a piece of glass he coated
> With collodion, and plunged it
> In a bath of lunar caustic
> Carefully dissolved in water—
> There he left it certain minutes.
>
> Secondly, my Hiawatha
> Made with cunning hand a mixture
> Of the acid pyrro-gallic,
> And of glacial-acetic,
> And of alcohol and water
> This developed all the picture.
>
> Finally, he fixed each picture
> With a saturate solution
> Which was made of hyposulphite
> Which, again, was made of soda.
> (Very difficult the name is
> For a metre like the present
> But periphrasis has done it.)

For the studio photographer this labor-intensive process might
have been tedious, but was not too difficult to undertake in a dark-
room with plenty of space to lay out the equipment. Out in the middle
of nowhere, on horseback, it was a different matter. To capture landscapes Muybridge needed a studio on the spot, wherever he was operating. Within a few months of arriving back in San Francisco, he had
devised and commissioned a box-shape wagon that he called his "Flying Studio." This was a small, two-wheeled carriage pulled by a single
horse. The upper half was covered in canvas, while the lower wooden
portion proudly had the words "The Flying Studio" painted in an arch
above his logo.

With this dynamic piece of kit came a new trade name. Perhaps he
felt that "Eadweard Muybridge" sounded too much like a photographer firmly based in an old-fashioned gallery. For the Flying Studio
business he devised the brand name Helios, after the Greek god of the
Sun. This was not just his trademark but his pseudonym as well; so
photographs would be "by Helios," which did not seem too strange at a

time when Dickens was writing under the equally enigmatic name Boz. The back of the Flying Studio and his business cards proudly displayed the Helios logo of a camera with wings, basking in the rays of the Sun.

Over the next few years Muybridge was to take hundreds of photographs of San Francisco, often in the popular stereogram format. As well as straightforward views of the better-known buildings and sites he indulged in romantic variants, most often producing "moonlight" views, which were perfectly ordinary photographs dimly exposed with an obviously artificial moon introduced by stopping the development of the image in a circular patch of sky using a piece of card to prevent the light rays from getting through to complete the exposure.

But just as he had first intended when taking his fateful visit to Europe, it was the remarkable views he took of Yosemite and other wilderness areas around California that distinguished Muybridge's early photographic work. His ventures into untamed country would justify his experience with on-the-spot work in the Flying Studio a hundred times over. The 150-mile journey required just as arduous an expedition as the worst parts of the overland trip from the East Coast. Some indication of what it took can be gathered from an article on Muybridge's trip published in the November 1869 edition of *The Philadelphia Photographer*. According to the writer:

> To photograph in such a place is not ordinary work. It differs somewhat from spending a few hours with the camera in Fairmount or Central Park. All the traps, and appliances, and chemicals and stores, and provender, have to be got together, and then pack mules secured to carry the load, and drivers to have charge of them. Thus accoutered, the photographer sets out, say, from San Francisco, through hill and vale, across deep fords, over rugged rocks, down steep inclines, and up gorgeous heights, for a journey of one hundred and fifty miles. Several days are thus occupied, and several nights of rest are needed along the road.

From his friend Selleck's gallery, Muybridge issued a first set of pictures of Yosemite. A brochure commented that "these views are by HELIOS and are justly celebrated as being the most artistic and remarkable photographs ever produced on this coast." His were not the first photographs of Yosemite, but his talent transformed the class of the pictures. Where the early photographs had given a very flat sense of

Muybridge's Helios Flying Studio in Yosemite.

a small view onto something vast, the Muybridge images captured the scale, the wonder, and the vitality of the wilderness.

This was certainly the opinion of the newspapers that reviewed Muybridge's work. The *San Francisco Daily Morning Call* of February 17, 1868, enthused:

> The views surpass, in artistic excellence, anything that has yet been published in San Francisco, combining, as they do, the absolute correctness of a good sun picture after nature, with the judicious selection of time, atmospheric conditions and fortunate points of view. In some of the series we have just such cloud effects as we see in nature of oil-painting, but almost never in a photograph.

The reference to "sun pictures"—really just an old term for a photograph, a picture made by the sunlight—might have been inspired by Muybridge's pseudonym Helios, the ancient Greek Sun god. This certainly seems to have been the case in another review from a few days earlier in the *San Francisco Bulletin,* which commented that a new Yosemite series had been taken by a photographer of the city who hid his name under the "significant classicism of 'Helios.'"

The same review tells us that there were "effects in some of the new views which we have not seen before." It praises the way the "plunging movement and half vapory look of cataracts leaping 1,000 or 1,500 feet at a bound" were wonderfully realized. This encouraging write-up was to be put in the shade by the paroxysms of delight experienced by the *Daily Alta California* on February 19th:

> [Helios] presents the gorgeous scenes of that locality from points of view entirely different from any heretofore taken. The series consists of twenty pictures showing the grand features of this wild and wonderful piece of Nature's architecture under the varying effects of the ever changing atmosphere.

What the reviews all pick out, quite surprisingly given his relatively recent start as a professional photographer, is how far Muybridge was prepared to go to get the right picture, fighting his way to locations that were more spectacular that the usual, easy, "point-and-shoot" locations that had been used before. And they remarked on how he captured the whole atmosphere of the place, not just a static rock formation or stand of trees. Later in the same article are some specific, almost poetic descriptions. Many of the Yosemite features had yet to be given anglicized names. They kept their older American Indian names that seem to give them something extra, even if the attempt to phoneticize the language with hyphens left some odd-looking words:

> In one picture you have before you the long vista of the gorge stretching away in magnificent perspective, palisaded on either side by cliffs, until lost among the mountains that hang like shadows on the distant back-ground. One of the finest of these perspective views is a piece called "The Mariposa Trail" embracing within its scope the lofty precipices and cascades of Tissa-ack, Tu-tuch-ah-no-lah and Pohono. It seems as though the artist had arrested the descending sheet of water until its mottled and foamy surface had paid tribute to his genius. . . . We are sure that the five months of assiduous labor spent by "Helios" in that place in securing a collection of

views that is unsurpassed by any preceding, will find an abundant return in our appreciative community.

Despite the effusive praise, Muybridge had not burnt his boats and committed himself entirely to scenic work. The same brochure that described the Yosemite photographs informed potential customers that he was willing to accept commissions to photograph "Private Residences, Views, Animals, Ships etc. anywhere in the city or any portion of the Pacific Coast." The photographer was named as Helios, but care of Edw. J. Muybridge. It is interesting the way Muybridge distances himself from Helios, even though his name appears twice on the page.

The photographs were a significant success and triggered a succession of commissions, most notably an expedition to Alaska, newly bought from the Russians. The U.S. Secretary of State at the time, William Seward, was suffering public humiliation at the time for wasting money on what was seen as wasteland with no redeeming value. Part of the purpose of Muybridge's trip was to provide a graphic demonstration of the potential of this vast new state. His immediate employer, Major General Halleck, was impressed with the result:

> I have to acknowledge the receipt of copies of your photographs of the forts and public buildings at Sitka and other military posts, taken for the use of the War Department, and also views of scenery in Alaska. These views besides being beautiful works of art give a more correct idea of Alaska, its scenery and vegetation than can be obtained from any written description of that country.

It's not often that a major general compliments one of his suppliers on producing beautiful works of art. Other, less high-profile commissions, though still with plenty of dramatic potential, took him along the length of the Pacific coast, photographing lighthouses and along the newly built Pacific railroads. Muybridge had become a name to reckon with in landscape photography.

The first trip to Yosemite in 1867 produced passable results, but Muybridge realized that the technology he was using was limiting his ability to put across the sheer scale of his subject. At the time, photographic prints were largely made by contact—the negative glass plate was put directly onto the chemical-soaked paper, then a light was shone through the plate, resulting in a picture that was the same size as the negative itself (in practice slightly smaller as the image did not entirely fill the negative plate), based on a sheet of glass that was typically only 8 by 6 inches.

The photographic paper used at the time was almost universally based on egg white (albumen). The albumen was whisked up with salt water, then used to coat one surface of a sheet of paper. The paper was dried and stored for use, when it would be floated on a bath of silver nitrate, dried again, and exposed through a glass plate carrying the negative image. The effect of the egg white was to hold the silver nitrate in a layer of constant consistency, to keep it from working its way unevenly into the fibers of the paper. (An unexpected side effect of albumen paper was a sudden surplus of egg yolks, millions of which were left behind as a by-product of the process. The manufacturers of photographic paper usually made arrangements with local industrial catering concerns that could make practical use of this large-scale outflow of yolk.)

As prints were now being made directly onto this albumen paper, they presented an obvious solution to the problem of getting bigger images—and one familiar to any modern photographer (at least before the advent of digital cameras)—an enlarger. The idea is simple enough. Instead of placing the negative right on the paper to produce a contact print, use a lens to project an enlarged image onto the paper. This was hardly an amazing concept even for the early days of photography, as the enlarger is nothing more than a magic lantern. These early versions were crude projectors—in effect, a camera obscura focused on a nearby picture—that had been used for many years to project a large image on a wall or sheet for entertainment.

Remarkably, an early form of magic lantern is described in a book dating back to the fifteenth century. In his *Liber Instrumentorum* (Book of Instruments), Giovanni de Fontana describes a strange device that projected a diabolical image on the wall: "a nocturnal appearance for terrifying viewers." This early spook show was actually in the form of a lantern with no lens—little more than an easy way of casting shadow pictures.

No one knows for sure when someone brought in a lens, but the Dutch optical scientist Christiaan Huygens was certainly building some sort of magic lanterns by 1659 (his father wrote to him, asking to be sent one so he could "frighten his friends" with it; the early lantern seems to have been a practical joker's favorite). By the mid-1660s, Thomas Walgensten had coined the term *laterna magica* for this type of device. By now the lantern, with its protruding lens structure, was more like a bull's eye lantern than an old-fashioned hanging lamp. Initially, the projections consisted of line drawings and then transparent paintings, some tinted in color, but the introduction of photography gave the magic lantern a much more realistic and impressive source for its images.

So why didn't the magic lantern immediately transform into the enlarger to produce bigger photographs? There were two problems—optical quality and grain size. It didn't really matter, when projecting a magic lantern show, whether the image was exactly even. Although high-quality lenses had been developed for cameras by Muybridge's time, the larger lenses of the magic lantern were usually of relatively poor specification. More important, though, was the grain size.

All the enlarger's lens can do is to make the grains in the original negative bigger and bigger. Eventually the individual grains become clearly visible as a series of shaded blobs. The result is a "grainy" photograph. When using modern films, which have very even and extremely tiny grains, this isn't a problem, though some photographers intentionally use grainy stock for effect. But the grains in early photographic materials were highly variable in size, and easily turned a photograph into a blotchy mess when magnified.

Although photographic enlargers had first appeared in the United States in 1843, they were still producing unsatisfactory results in the 1860s and 1870s. Muybridge wasn't happy with the resultant distortion. He wanted to produce a crisp, large image that would rival a painting as a work of art to be displayed on the wall of a home, and that would do justice to the scale and grandeur of Yosemite. His triumphant return to the wilderness in 1872 would do just that. The plates Muybridge used were up to 24 by 20 inches in size—huge negatives that cemented his growing reputation as one of the outstanding landscape photographers of his time. It was the stunning photographs Muybridge took on these mammoth plates that would win him the

International Gold Medal for Landscape at the 1873 Vienna Exhibition. It also made his trek out into Yosemite even more of a feat. He now required not only packs containing these huge and heavy pieces of glass but also much bigger tanks and boxes for the development process and for storing the oversized negatives.

On his second trip to Yosemite, Muybridge was anything but a passive observer of the landscape, nor was he prepared to stick to well-trodden paths. In the *Alta California* of April 7, 1872, we hear that

> he has waited several days in a neighborhood to get the proper conditions of atmosphere; for some of his views he has cut down trees by the score that interfered with the cameras from the best point of sight; he had himself lowered by ropes down precipices to establish his instruments in places where the full beauty of the object to be photographed could be transferred to the negative; he has gone to points where his packers refused to follow him, and he has carried the equipment himself rather than to forego the picture on which he has set his mind.

Just as he had done with his book trade, Muybridge was by now trying to get subscribers for his photographic works. The outlay for equipment and materials was large, and any guaranteed income would help defray the costs. To this end he put out a prospectus for his expedition. "At the suggestion of several artists and patrons of Art," it informs us, "I propose devoting the approaching season to the production of a series of large-size photographic negatives, illustrating the Yosemite and other grand picturesque portions of our coast."

Later in the prospectus he made it clear what this was all for:

> [S]ubscribers for each one hundred dollars subscribed will be entitled to select FORTY from the whole series, to be printed and mounted upon India tinted boards, in every respect similar to my smaller ones. It is scarcely necessary to say in consequence of the great expense attending the production they will not be sold at this price excepting to subscribers.

This was not a purchase for the financially challenged—that $100 would be the equivalent of around $2,000 now. It might seem that he was unlikely to get much return this way, but the copy of the prospectus in Muybridge's scrapbook has a list of subscribers that pushed the revenue to around $20,000—perhaps $400,000 in modern values. The photographic business was much more risky than bookselling, and the overhead much greater, but the potential reward was higher too.

In 1873, he published a catalogue of his wilderness images, including the earlier pictures and his latest additions, grandly titled *Catalogue of Photographic Views, Illustrating the Yosemite, Mammoth Trees, Geyser Springs, and Other Remarkable and Interesting Scenery of the Far West, by Muybridge.* In the catalogue Muybridge makes it clear how recently Yosemite had become a public wonder. He wrote that "it was discovered in 1851, is situate in the county of Maraposa, about 250 miles from San Francisco, and is about 4,000 feet above sea level. In 1864 it was granted by the United States Government to the State of California, for the free use and recreation of the public for all time to come." By the time this comprehensive collection was published, his business links with Selleck were well in the past. Surprisingly though, considering the strong entrepreneurial streak in the Muybridge makeup, the catalogue was not self-produced. The publisher is listed as Bradley and Rulofson, the leading San Francisco photographic studio of the day.

Muybridge had moved out from under Selleck's wing in 1869 to share premises with Charles and Arthur Nahl, two brothers at the heart of the San Francisco art circle. He stayed with them until 1873, when he moved on to Bradley and Rulofson. Muybridge had outgrown the Nahls. While Charles and Arthur Nahl had the artistic edge, the larger company had the resources to cope with his immense photographs, bringing in specialist printing equipment to deal with the gargantuan negatives. This was no corner shop. When Muybridge joined them, Bradley and Rulofson had 36 employees and a luxurious gallery equipped with the latest fittings, including one of the first elevators in San Francisco.

Despite the dubious benefit of the elevator—Muybridge was very wary of it and never used it if he could avoid it—there was no doubt of the benefits of working with Bradley and Rulofson, just as Muybridge had gained from his association with the Nahls. Muybridge had not become an employee—far from it. He usually acted as his own publisher and publicist. Yet the nature of his work, which often entailed weeks or months of travel, made it impractical to set up a studio operation, where inevitably much of the business revolved around portraits. Instead, Muybridge preferred a convenient association with a more traditional establishment. After his years in the book business he no longer wanted the overhead of keeping up his own business premises.

On the whole, Muybridge's relationship with Bradley and Rulofson was a good one, though there were teething troubles. Another gallery, belonging to a Thomas Houseworth, felt that they had some call on Muybridge's work. When the large-scale photographs of Yosemite were published, Houseworth's put a very poor, battered copy of one print, clearly identified as Bradley and Rulofson merchandise in their window, as if to indicate the indifferent quality of Muybridge's output when working with Houseworth's competitor. Muybridge appears to have found this amusing, as he kept some cuttings on the affair in his huge scrapbook, something he pointedly didn't do when he did not approve of the coverage (for example that of his murder trial).

The first entry in the saga was a shot fired across Houseworth's bows with a notice in the *San Francisco Daily Evening Bulletin*:

> Messrs. Bradley and Rulofson are much obliged to Mr. Houseworth for giving their names a place in his window; but attaching them to an old, soiled print from a condemned negative of Muybridge's (neither print nor negative being made by them), shows to what a wretched straight the poor gentleman is driven in a fruitless effort to compete in business.

Houseworth was not going to take this shot lying down. He quickly riposted with his own *Evening Bulletin* notice:

> To the public in general, and a reply to the card of Bradley and Rulofson— The Yosemite View exhibited by us in our window is one of a set of forty furnished to a subscriber by Bradley and Rulofson for the sum of $100 and bears their name as publishers. The View is a fair sample of the lot which was sold to me at a heavy discount on the cost and is now in the same condition as when received by the original purchaser. We would further remark that we had tried to purchase from these gentlemen some of their views and they positively refused to sell us, for reasons which we leave others to judge.

Just so there was no doubt about his personal judgment, Muybridge himself also entered the fray, publishing this cryptic but pointed response in the *Evening Bulletin*:

> Aesop in one of his fables related that a miserable little ass, stung with envy at the proud position the lion occupied in the estimation of the forest residents, seized on some shadowy pretext of following and braying after him

with the object of annoying and insulting him. The lion turning his head and observing from what a despicable source the noise proceeded, silently pursued his way, intent upon his own business, without honoring the ass with the slightest motion. Silence and contempt, says Aesop, are the best acknowledgements for the insults of those whom we despise.

While it's not entirely clear whether the lion is Muybridge himself or Bradley and Rulofson, it's pretty obvious who the "miserable little ass" is.

∞∞

Although the subject Muybridge most preferred was landscapes, as we have seen from his advertising, he was willing to take photographs of anything from animals to houses. While he specifically excluded formal portraits, he was not above photographing people, and the opportunity to bring human interest into his work came to the fore shortly after he had moved under the wing of Bradley and Rulofson.

Like Muybridge himself, his new associates took what would now be termed a proactive approach to business. They did not sit back and wait for commissions, but actively looked for new opportunities to take interesting photographs that they would subsequently sell. It was their habit, for example, to scan the newspapers for announcements of celebrity visitors to San Francisco, then to track them down and photograph them. In this way they acted like a photographic Reuters of their day, capturing news images for whoever would like to buy them. It was not entirely surprising, then, that Bradley and Rulofson should wish to sell photographs of a local skirmish that was to become known as the Modoc War, one of the last significant stands by American Indians against the encroachment of the European settlers.

By 1872, most of the American Indian tribes had moved beyond the intertribal battles that had once been common, sinking into a state of inertia. But the Modocs of northern California were preparing for one last struggle. They had been transplanted to the reservation of the Klamath in Oregon, a tribe they had little time for. A small band of Modocs, headed up by Kientpoos, who also went under the name of Captain Jack, returned in an attempt to reclaim their traditional hunting grounds based around Tule Lake in northern California (later the site of an infamous World War II internment camp for Japanese nationals) in an area known as the Lava Beds.

Attempts to get the Modoc band to return to the reservation in late 1872 and early 1873 resulted in routs for the American soldiers. They were in a classic position of trying to find and dig out a group of resistance fighters who were occupying difficult territory that they knew well. In one raid, from a troop of 400 soldiers, 35 were killed and many more injured without a single Modoc ever being seen.

Attempts were made to negotiate a settlement, but neither side seemed serious about offering anything of value. Any possibility of peace disappeared in a horrendous meeting on April 11, 1873, when both sides turned up with weapons at a supposedly unarmed truce talk. The American General Canby and another senior negotiator were killed, and total war was declared.

Muybridge was sent out to the Lava Beds soon after. He photographed not the fighting but rather the goings on behind the lines, from wounded soldiers to officers at the campaign table. He also made detailed studies of the territory, largely unknown from the U.S. side, which were forwarded to Washington to emphasize the difficulty under which the U.S. soldiers were operating. His pictures, converted into engravings for publication, were seen nationwide. (At the time photographs could not be reproduced in print.) Sometimes, like many photojournalists who followed him, he was tempted to adjust reality to suit the camera. He did not have access to the Modocs themselves except as corpses, yet he managed to send back a picture of a Modoc warrior, lying sniper fashion, ready to shoot the U.S. troops. In fact, Muybridge's subject for this photograph was a Tenino Indian acting as a scout for the U.S. Army, who agreed to pose for Muybridge.

The end came soon after. Following the disastrous earlier attempts, the U.S. troops were finally trained in the Indian form of warfare, getting a better understanding of the landscape and how to use it. Now their huge advantage in numbers could pay off instead of being an inconvenience. The remaining Modoc braves, less than a hundred men, were captured and dispatched to a reservation in Oklahoma. But Muybridge had already moved on.

FOUR

STANFORD, STONE, AND LARKYNS

The city of San Francisco is built along the eastern base and up the side of a row of high sand-hills, which stretch southwardly from the Golden Gate, between the Pacific Ocean on the west and the bay of San Francisco on the east. The city has been built out into the bay some fifty to a hundred rods by carting in sand from the eastern slope of the hills, which are thus left more abrupt than they originally were. The compactly built district seems rather more than two miles north and south, by somewhat less east and west. I judge that the city is destined to expand in the main southwardly, or along the bay, avoiding the steep ascent towards the west. The county covers 26,000 acres, of which one-half will probably be covered in time by buildings or country-seats. I estimate the population at about 80,000.

From *An Overland Journey from New York to San Francisco in
the Summer of 1859,* Horace Greeley

During his time in San Francisco, Eadweard Muybridge was to meet three individuals who would between them shape, and come close to destroying, his entire future—the young divorcee Flora Shallcross Stone, the dashing rogue Harry Larkyns, and the railroad magnate and university founder, Leland Stanford.

Stanford picked Muybridge out of the blossoming field of photographers in San Francisco to discuss using a camera to analyze the motion of a trotting horse. He imagined taking a photograph so quickly that it froze the position of the limbs and established whether the horse ever "flew" with all four legs off the ground simultaneously. The apocryphal story goes that Stanford needed the photograph to win a bet of

$25,000 (around half a million dollars in today's money), but as we shall discover, the reality was probably very different.

Muybridge declared himself "perfectly amazed at the boldness and originality of the proposition." Although many photographers had tried to photograph rapidly occurring action, none had managed to capture such a fleeting moment as a snapshot of a horse's gait. Stanford wanted to halt the movement of trotting horses, to dissect the individual components of their gait. At a time when photographs often took many seconds to expose, this feat was beyond the capabilities of the technology—a challenge indeed for the man who had come to be regarded as California's greatest landscape photographer.

Stanford was an entrepreneur turned politician and horse breeder. The pinnacle of his business career had been the building of the Central Pacific Railroad, the final, western half of the first railroad to cross the American continent from shore to shore. Stanford had not initiated the project, but had been president of the company since its inauguration in June 1861 (the line was finished in 1869), and had thrown all his considerable abilities into making the railroad a reality. Even his successful term as governor of California from 1862 to 1863 was tied strongly to his need to make the railroad a success. But heading up a major railroad project was no cushy office job, and Stanford was beginning to feel the strain. His doctor had told him to find an interest outside his work and he diverted part of his undoubted energy to horses.

Following the vogue for applying scientific enquiry to every aspect of the world, the racing fraternity had developed a fascination with the mechanics of motion. Breeders and trainers longed to analyze the way a horse's legs propelled it along—to understand the motion of its limbs in the different carriage modes of the horse, in the hope that understanding would lead to improvement. In Stanford's case the particular interest was the trotting race, originally a normal horseback race where the horses were limited to a trot; but by the 1830s, it had been transformed into a harness race, where the trotting horses pull lightweight, skeletal carriages along the track. With this new style, trotting races had become big business.

The horse's distinct gaits arise from its four-legged layout. Wild

horses normally use only two gaits—the walk and the gallop—but domestic breeding has introduced two more ordinary gaits, plus a few
specialties limited to a subset of horses. The four familiar gaits of the
modern domesticated horse are the walk, the trot, the canter, and the
gallop.

In a walk each of the horse's feet takes a separate step, one after the
other. Like almost all four-legged mammals, the sequence of steps is
left hind, left front, right hind, right front. Although these paces are
taken in an even way, it wouldn't be possible to wait for the left hind
foot to hit the ground before starting to move the next foot, so the
animal sways between having three feet and two feet on the ground.

When moving up to a trot, the horse combines diagonally opposite pairs of feet, taking its weight on, say, the right front and left hind
legs while it moves the other two into position, then shifting its weight
to the left front and right hind legs. For the canter or lope, one of the
horse's front legs moves first, followed by the other foreleg and its diagonally opposite rear, then the remaining rear leg, producing a strange
three-beat gait that is asymmetrical, depending on which foreleg leads
off. Usually either one or two of the feet are on the ground, but in the
canter there is an audible pause after the third beat of the hooves, when
all feet are away from the ground.

Finally comes the gallop. Here the horse is running full out. It's
effectively an extended canter where it's just not possible for the two
legs in the middle of the gait to operate together, so the hind leg of the
pair touches down just before the foreleg, going back to four beats like
a walk (though of course much faster). Again there is a period when all
four feet are off the ground, slightly longer than in the canter and hence
even more obvious.

Although it wasn't clearly understood then, the "bounce" that takes
four feet off the ground in the faster gaits from trot to gallop is an
essential part of the horse's efficiency in motion. It has been said in the
past that kangaroos should not exist, because it's impossible to produce enough energy to move the way they do. If you sum the energy
required to produce each of a kangaroo's bounces in a day, it appears
to use up more energy than it takes in as food—an impossibility. What
this analysis overlooks is the efficiency of the bounce process, where

much of the energy needed to take off is given by a springy absorption of the energy with which the animal hits the ground at the end of the previous bound.

Think of a rubber ball. Say you drop it from waist height and it bounces 20 times. Add up the energy required to lift it up in the air those 20 times and it will be much more than the energy you originally gave it by bringing it up to your waist. Similarly with the kangaroo's elastic progress, and also as recent research has shown, the horse, in its "flying" gaits, has enough springiness in its bounce to seemingly put out more energy than it has absorbed in food.

Stanford, as we have seen, was interested in trotters. He reckoned that the more he could know about their gait, the better he could breed for certain characteristics. He had already done experiments to determine where the horse put most effort in its gait by trotting his livestock across a smooth bed of sand and examining the indentations, equating a deep indentation with greater effort (and hence a need to build greater muscular strength). Yet there was another aspect of the trot that no one had successfully explored. As we have seen, the trot involves alternate diagonal pairs of legs supporting the horse—but what happens at the changeover? Either there's an overlap where all feet touch the ground, or a moment of liftoff where all four feet are in the air at once; in principle, whichever was the case would require a different muscular effort on the part of the horse that might need subtly different training procedures. So the flying horse wager was born.

In this story, history and myth became intertwined. The wager, usually said to be between Stanford and either James R. Keene or Fredrick MacCrellish, has been described elsewhere as fact. It certainly was mentioned in newspaper articles of the time, but none of the parties involved ever substantiated it. In fact, the chances are high that this was a story dreamed up tabloid style in a quiet period to add a little spice to the newspapers. As Janet Leigh, an ardent Muybridge researcher in the 1940s and 1950s put it in a letter, it was a piece of "postmortem press agent work." Leigh argues that it would have been totally out of character for Stanford, who was never known to gamble, to take part in a wager, and had there been a bet, it certainly wouldn't have been with Fred MacCrellish, the proprietor of the *Alta California* newspaper, who took the same side in the argument as Stanford (both believed that a horse's feet did all leave the ground simultaneously when it moved at speed). Their opponents were Robert Bonner of the New York *Ledger*, George Wilkes of the New York *Spirit of the Times*, and James R. Keene, president of the San Francisco Stock Exchange.

As such, Keene perhaps could be seen as a likely gambler—a stock exchange is little more than a highly formalized betting mechanism—yet mere involvement in the gambling business does not make you a gambler. The lack of direct evidence plus the known character and attitude of the parties involved suggest that Mrs. Leigh was right, and the bet was simply concocted by the papers to give color to what was undoubtedly a hotly disputed issue.

Finding a way to prove the outcome one way or another had been occupying scientific minds for a number of years. The French physician Étienne-Jules Marey had tried to plot the motion of a horse's feet by using pressure plates to change the pattern of a line drawn on a revolving cylinder. Each of the horse's feet was fitted with a separate shoe with a squashable leather chamber at the bottom. As the hoof hit the ground, the plate compressed the chamber and sent a blast of air up small pipes leading to a recording device on the horse's back that translated the variations in pneumatic pressure into movements of a pen on a roll of paper. But this method had not given enough information to decode the dynamics of a fast-moving horse. The mechanical system could simply not respond fast enough, nor be precise about what was happening on the ground.

According to Muybridge, it was MacCrellish who suggested that Stanford should use Muybridge and his photographic skill to settle the issue once and for all. This was contradicted by A. C. Rulofson, the son of Muybridge's business associate William Rulofson. Rulofson junior claimed that Stanford approached Bradley and Rulofson first, and it was in their studios that "Stanford explained to them his ideas for a number of cameras." According to A. C. Rulofson, Muybridge got the job as the most experienced outdoor photographer known to the firm. However, this story is a trifle suspicious; there is no particular reason for Muybridge to lie about the source of the contact, while Rulofson may have wanted to give his father an inflated role. And there was no suggestion of using multiple cameras in Muybridge's early experiments, so at least some of this report seems to come after the fact.

At the time, Stanford was living in Sacramento, around 75 miles inland from San Francisco. Although already a great equestrian enthusiast, Stanford did not have his own stables; he kept his horses at the Union Park racetrack. It was there that Muybridge first encountered Occident, Stanford's favorite and the subject of their early attempts at stopping time with the camera.

Occident was a real life Black Beauty, rescued after being used cruelly by a sequence of owners. One liveryman had a habit of firing his gun between the horse's ears in an attempt to dampen his spirit, while another had put the lightweight animal to work dragging great earth-moving carts. When Occident proved incapable of heavy cart-horse work he was sold again.

The next owner treated Occident better, and even entered him in a scratch race, which to everyone's surprise the novice horse won. There was now only one further home for Occident before he came to Stanford, from a trainer called Eldred for whom Occident had already shown some style, though not enough to warrant the $4,000 that Stanford paid for the horse. Eldred assumed that Stanford was a typical amateur with more money than expertise, but Stanford was not a man to take anything lightly. He had gone into the horse business with a combination of entrepreneurial verve and scientific understanding, and he was convinced that he had found a winner in Occident. History proved him right.

Within a couple of years, Occident was known throughout the country as a remarkable trotter, though he was not to enter his first big race, which he lost, until October 1872. It was the next year that his career really took off, by which time Muybridge had already made his first attempt at freezing Occident's motion on a photographic plate.

Compared with the later, better documented trials, when Muybridge would achieve remarkable things, the exact details of this early experiment in instantaneous photography made in 1872 are uncertain. We know that the attempt was made in May of this formative year—the same year as Eadweard's marriage and his triumphant expedition to Yosemite—but no photograph has survived.

Some believe, however, that Muybridge did succeed in showing Occident with all four feet off the ground, even though at the time doubts were expressed about the achievement. The challenges were immense. To freeze motion required an exposure of a fraction of a second, at a time when typical shutter technology involved the removal of a hat held over the lens, counting seconds aloud slowly for anything up to a half a minute, then putting the hat back. What's more, instantaneous photography required a plate that was sensitive enough to record an image in such an extremely short span of time, when most exposures took seconds. This long exposure time simply reflected the time taken for the silver salt grains on the wet collodion plate to produce a reasonable image.

In the move to instantaneous photography no one person was responsible for a technical breakthrough; instead, improvements involved repeated tweaking of the technology, with word spreading through the photographic community. One essential was better lenses, which were available by the 1870s, notably those from Dallmeyer in London. These were designed to maximize the amount of light transmitted through to the plate. But most of the development was in the effectiveness of the chemical process. Different mixes of silver salts, different temperatures, variations in the time taken for each stage of the preparation process—each was modified, sometimes simply by fortuitous accident, until the plates were capable of producing a result, however lacking in contrast, in as little as a hundredth of a second.

Muybridge admitted that the photographs he took at the time were little more than a silhouette, and as was normal practice, the images he showed the public were heavily retouched. It would have been easy to have proved anything, had the perpetrator an axe to grind. There was no immediate report in a newspaper, but oddly there was a story the following year in the April 7, 1873, edition of the *Alta California*, the paper of Fred MacCrellish, in Stanford's camp, written as if it were a contemporary account. It may have described a second attempt, but it is equally likely that the reporter was simply making the story more dramatic by suggesting it had just happened. There is certainly some sign of an attempt at dramatization in the description of initial failure:

[Stanford] wanted his friends abroad to participate with him in the contemplation of the trotter "in action," but did not exactly see how he was to accomplish it until a friend suggested that Mr. E. J. Muybridge be employed to photograph the animal while trotting. No sooner said than done, Mr. Muybridge was sent for and commissioned to execute the task, though the artist said he believed it to be impossible; still he would make the effort.

All the sheets in the neighborhood of the stable were procured to make a white ground to reflect the object and "Occident" was after a while trained to go over the white cloth without flinching; then came the question how could an impression be transfixed of a body moving at the rate of thirty-eight feet in a second.

The first experiment of opening and closing the camera on the first day left no result; the second day, with increased velocity in opening and closing, a shadow was caught. On the third day Mr. Muybridge, having studied the matter thoroughly, contrived to have two boards slip past each other by touching a spring and in so doing to leave an eighth of an inch opening for the five-hundredth part of a second, as the horse passed, and by an arrangement of double lenses, crossed, secured a negative that shows "Occident" in full motion—a perfect likeness of the celebrated horse. The space of time was so small that the wheels of the sulky were caught as if they were not in motion. This is considered a great triumph as a curiosity in photography—a horse's picture taken while going thirty-eight feet in a second!

As no one claimed anything more than a silhouette, the article's description of a perfect likeness suggests significant journalistic license. Perhaps a more accurate account was Muybridge's own, where there is no dramatic buildup; but it was written a quarter of a century later for a book entitled *Animal Locomotion*, in which Muybridge set the scene for the experiment to discover whether the horse had all four feet off the ground at any time.

The attention of the author was directed to this controversy, and he immediately resolved to attempt its settlement. The problem before him was to obtain a sufficiently well-developed and contrasted image on a wet collodion plate, after an exposure of so brief a duration that a horse's foot, moving with a velocity of more than thirty yards in a second of time, should be photographed with its outlines practically sharp. . . .

Having constructed some special exposing apparatus and bestowed more than the usual care in the preparation of the materials he was accustomed to use for ordinary quick work, the author commenced his investigation on the racetrack at Sacramento, California, in May, 1872, where he in a few days made several negatives of a celebrated horse named Occident, while trotting, laterally, in front of his camera, at rates of speed varying from two minutes and twenty-five seconds to two minutes and eighteen seconds per mile.

The photographs resulting from this experiment were sufficiently sharp to give a recognizable silhouette portrait of the driver, and some of them exhibited the horse with all four of his feet clearly lifted, at the same time, above the surface of the ground.

So far as the immediate point of issue was concerned, the object of the experiment was accomplished, and the question settled for once and for all time in favour of those who argued for a period of unsupported transit.

That "unsupported transit" is having all four feet off the ground at once.

The odd speed measurement of "two minutes and eighteen seconds per mile" is one that is more familiar from athletics, where we might meet anything from a four-minute mile to 100 meters run in 9.79 seconds. Occident's fastest rate is the equivalent of 26 miles per hour. It seems that such a description of speed was unusual even at the time, at least outside the United States, as it confused contemporary observers abroad. A number of years later, one W. H. Davies, lecturing at the Edinburgh Photographic Society showed a selection of Muybridge's photographs. Seeing a speed given as 2:40, Davies made a rather thoughtless comment, saying "I may mention the general speed of a fast trotting horse is about two and a half miles a minute." This would have them traveling at 150 miles an hour. In response, *Scientific American* remarked sarcastically in February 1880: "Our cousins across the water are so unaccustomed to the sight of fast trotters that perhaps it is not surprising that the lecturer's statement should have been received as correct."

Whatever the outcome of the experiments, there is no doubt that Leland Stanford's attempts to understand the horse's gait would have a profound and positive effect on Muybridge's future.

☙❧

Flora Stone's influence was a very different kettle of fish. Muybridge met his wife-to-be through the Nahl brothers. Charles Nahl was largely a painter, specializing in huge life scenes. Arthur, though, was thought to be one of America's best photographic retouchers. This was a craft that had sprung from nothing to become big business. The photographic images of the time were often flawed because of the difficulty of spreading the chemical mix evenly on the glass plate. And anyway, particularly in portraiture, clients were used to seeing a picture the way they wanted it to appear, rather than relying on the more naturalistic approach we might take today. Retouching was in part the pure mechanics of removing the specks left by dust and the distortion of the collodion gel but also involved the art of subtly transforming the appearance of the sitter, or painting in colors to tint a photograph and make it look more natural.

Just as an artist might in practice paint only a small proportion of his studio's works directly, directing the efforts of lesser assistants in something close to a production line, so the best photographic retouchers did not sully their hands with everyday work. Their assistants, often women, handled the repetitive detail, leaving the more difficult detail to the masters. One such assistant was Flora Stone.

At the time, Flora was still married to Lucius Stone, who imported, made, and sold saddlery. But the marriage had already irretrievably broken down. Flora alleged problems due to Stone's age and the cruelty of his mother, with whom the couple lived and who was Stone's business partner. It was probably after her separation from Stone that Flora went to work with the Nahls. She obtained her divorce shortly before she and Muybridge were married in 1872 (at least it is assumed that they were married, though no record of the wedding has ever been found).

By the time of her divorce, Flora was no longer working as a retoucher. She had moved on to the position of a shopgirl at the less prestigious establishment of Ackerman's Dollar Store on Kearny Street. She and Muybridge were very different, in both background and in age. Rather strangely, given the reason cited in her divorce, Flora was once again to marry a man much older than she—Muybridge was twice her 21 years. Where Muybridge had strong if physically disparate family ties, Flora's background was less conventional.

Flora was an orphan, originally from Alabama. She had lived with grandparents and an aunt, but as a teenager had become the foster daughter of Captain W. D. Shallcross. He paid for her schooling and called her Lily. We don't know anything about the relationship between

Flora and the captain, but we do know that she ran away to marry Stone without her foster father's consent.

Only three photographs of Flora have been identified with any confidence, and even these could, from their appearance, be the photographs of three different women. The first, taken when Flora (if it is truly she) was about 17, shows a rather plain young woman, with close-set, piggy, small eyes. The latest of the three is supposedly Flora while pregnant in 1874. This one shows a pinch-faced, flat-chested, dowdy woman, uncomfortably reminiscent of the cut-out animations Terry Gilliam provided for Monty Python's Flying Circus. It's tempting to expect her top half to open up, as if hinged, and the baby to pop out.

The intermediate photograph, dated to around 1872, is a very different affair. This Flora, sporting what must have been a fashionable hairstyle, stares brazenly over her left shoulder. Her figure is voluptuous, enhanced by a (for the period) low-cut, frilled top. It could be the work of those retouchers, but her eyes are now much larger, slightly hooded. Her mouth is downturned—she is not smiling in any of the photographs, though this wasn't uncommon at the time—even so, her expression suggests a faint, sophisticated amusement.

Although only these three photographs are acknowledged in much of the Muybridge literature, there is, in fact, a fourth one, clearly from the same series as the 1872 portrait, which was entered by Bradley and Rulofson in the *Philadelphia Photographer*'s prize competition in 1874. It won the medal, the judges declaring that the six copies of the photograph were "among the purest specimens of photography it has ever been our good fortune to inspect." In this version of the photograph Flora sits straight up, her face in profile, more demure than in the challenging photograph that is usually shown, but even so with a hint of fascination that might have contributed to its winning the medal. This is the Flora, demure yet knowing, who would have captivated the practical but unworldly Muybridge. This is the Flora who would bring him to murder.

Together they set up home on the edge of South Park, a slightly faded but elegantly designed suburb of San Francisco. By now Muybridge had a certain fame and was well off, despite a tendency to use up large amounts of money on his photographic expeditions. He

Flora Muybridge.

was also, in all probability, a rather boring, work-obsessed figure to someone like Flora. A photograph of Muybridge taken at around the same time as Flora's shows a gray or white-haired man with a grizzled beard. (He would later claim that his hair went gray overnight when he had his stagecoach accident, but this is highly unlikely.) This is not an old man, nor a decrepit one—in fact his eyes show a powerful energy and he is known to have kept himself fit—yet he is very clearly middle aged.

It was shortly after moving in at South Park that the Muybridges would be introduced to their nemesis, the third of the forces that changed Eadweard Muybridge's life forever, Major Harry Larkyns.

Larkyns was a man with a past, though how much of it existed only in his imagination will never be certain. He claimed to be related to the British royal family and to have become the favorite of the rajah of a distant Indian province. Larkyns boasted that he had succeeded the rajah when he died, inheriting both his throne and his harem until the sybaritic life become too tiresome. Some of his claims were better substantiated. Larkyns certainly fought with the French army in the Franco-Prussian War under the disastrous General Bourbaki, where he reached the rank of major. And he was unusually well traveled. He had journeyed to the four corners of the world, a restless existence that brought him to San Francisco in November 1872.

Larkyns was over 6 feet tall, towering over many of his contemporaries. Those who had met him described Harry Larkyns as dashing and handsome, a lady's man, and an adventurer. According to one source, he "always wore a Prince Albert coat, starched white linen and a black bow tie." He was known as a writer, a poet, and a musician. He was said to be a crack shot, a gifted scientist, and an expert cook. He was a charismatic charmer, always prepared to use his personal magnetism to his advantage.

Nothing illustrates Larkyns's charm better than the story of his first few months in San Francisco. Larkyns arrived in the company of a young man, Arthur Neil, with whom he had met up in Salt Lake City. Both were traveling to San Francisco, and Neil quickly found Larkyns's company fascinating.

Larkyns told Neil that by some blunder the express company had managed to send all his things—even most of his money—on to San Francisco ahead of him. As Larkyns was on his way to Yokohama in Japan, it would be almost impossible to keep track of his effects. He asked Neil if he could help out financially, just until they got to San Francisco, which the younger man was very willing to do. To reassure his new friend, Larkyns made regular reference to a letter of credit for £1,000 sent on by his rich English grandmother, which would be waiting for him in the city.

When the pair reached San Francisco, Larkyns suggested that they stay at the opulent Occidental Hotel. Payment, he assured Neil, would be no problem once he had managed to untangle his letter of credit. Unfortunately for the moment, though, his money had been accidentally sent on to Yokohama ahead of him, so perhaps Neil could oblige again by dealing with the bills. Larkyns proved to have a phenomenal talent for spending money. In the first week, his wine bill alone amounted to $127.50—the equivalent of around $2,500 now. Obligingly, Neil paid up.

Larkyns must have thought that he was in paradise. Whatever he wanted, he could afford—at Neil's expense. He spent vigorously on a young lady by the name of Fanny, a woman that the *San Francisco Chronicle* would later describe as notorious. He invited himself along when Neil took a trip to Honolulu and lived the high life there (at Neil's expense, naturally).

After three months of this treatment, Neil's patience finally snapped. He was tired of hearing variants of the same old story, of waiting for the money that was always just about to arrive. One morning in early March 1873, he lodged a complaint against Larkyns, accusing him of obtaining money under false pretences. By midafternoon Larkyns found himself in the San Francisco City Prison, awaiting trial. The *San Francisco Chronicle* made a meal of Larkyns's story, commenting:

> Let his fate serve as a warning to such young gentlemen amongst us who may be living on the credit of expected remittances.

Larkyns did not have long to wait to learn of his fate. Either Neil had pulled a few strings or the courts were quiet that week; Larkyns found himself put up for trial the next day. Yet even against the threat of imprisonment his remarkable charm won through. Larkyns persuaded Neil to see him briefly before the hearing. After just a few minutes, Neil dropped the charges, on the proviso that Larkyns stayed out of Neil's life in the future.

By the time the Muybridges were introduced to Major Larkyns, a little over a year into their marriage, the one-time conman appeared to have turned over a new leaf. He had got himself a steady job as a journalist, working as drama critic of the *San Francisco Post* and as business correspondent to the *San Francisco Stock Report*. Even so, Muybridge took a dislike to him from the start. He would later describe their early acquaintance in a newspaper interview:

> In the early part of 1873 he came into the gallery when I was at work. Mr. Rulofson and Max Burkhardt, who were employed in the gallery, had known him for some time previous to this. He wanted to get some of my

views for some purpose he had in mind, I do not now remember what. My
wife, who sometimes worked in the gallery a little, touching up pictures,
was present, and she introduced him to me. I had frequently heard her
speak of Major Larkyns before, but did not know that she was much acquainted with him—but it seems they were better acquainted than I expected they were.

She afterwards told me that she was introduced to him at the house of Mr.
Selleck, a produce dealer who resides at South Park, with whom I am acquainted. After his introduction to me he frequently came up to the gallery, and I often gave him points in regard to art matters which he was then
writing about for the Post. He was under such obligations to me frequently,
and as a return was always trying to get me to take passes to the theater, of
which he said he had plenty. I didn't want to accept them from him. In fact,
I am not much of a man to take favors from anybody. I seldom go to the
theater, and when I do I am always willing to pay for it. Besides, I did not
fancy the Major's style of man, and did not feel like placing myself under
obligation to him.

Larkyns has sometimes since been portrayed as a family friend of
the Muybridges', a man that Eadweard Muybridge trusted to accompany his wife to the theater and keep her entertained in his absence,
but it is clear that from the start Muybridge found the man distasteful.
He went on to describe what must have been the closest social engagement he had with Larkyns:

> [H]e insisted so often and so strongly that one day I accepted passes from
> him, and in the evening myself and wife went with him to the California
> Theater. After the play he and I had a drink of whiskey together and he
> accompanied us home. . . . I did not know or suspect that he was visiting
> my wife or even went anywhere with her. But one night, while Neilson
> [probably the popular Victorian actress Adelaide Neilson] was playing her
> engagement here, my wife was out when I got home, and did not come in
> until late. I asked her where she had been and she said, "To the theater with
> Major Larkyns."
>
> I then said earnestly, "Now look here, Flo, I don't want you to be going out
> with any man at night without my knowledge or consent. It is not proper
> and will bring you into scandal if you persist in it. Now, do you never go
> out with him in that way again." She promised me solemnly and faithfully
> that she would not.

Muybridge was not going to leave it at that, though. As might have
been expected of any husband of the time, he also put Larkyns in his
place:

The next morning I went to see Larkyns about it. I met him on Montgomery Street, just coming down from his rooms, and he said quite gaily, "Good morning, Mr. Muybridge."

I said "I hear that you were at the theater with my wife last night."

"Yes," said he, "but I didn't think there was any harm in that."

I said "You know very well that, as a married woman, it is not proper to her to be running about at night with you or with any other man but me, and I want you to let her alone. Do you never take her out in that way again. You know my right to speak in this manner. You know my right in the premises as a married man. So do I, and I shall defend them. If you transgress them after this morning I shall hold you to the consequences, and I suppose you know what that means in California."

"Oh yes," said he, "and if there is any objection on your part to her going out with me, I will take her out no more."

Larkyns was indubitably smooth-tongued, as Arthur Neil could testify, and had an abiding affection for the good life that would have proved attractive to Flora. For that matter, though Flora was young, she was already more worldly than her husband. Flora Shallcross Stone was a divorcee when Eadweard married her in his first and only marriage. There is no evidence that Flora was seduced by Larkyns—it is very likely she was just as willing as he was to make the most of their opportunity to have fun together.

Muybridge frequently left Flora on her own while he was off on photographic excursions. After little Floredo, their son, was born, though, he did not like to leave her alone with the baby. When, in the autumn of 1874 he expected to go off on a lengthy tour photographing the coast, since Flora had no relatives in San Francisco, he suggested she go up to stay with her uncle, Captain Stump, in The Dalles, Oregon. Muybridge would later deny that his moving Flora out of San Francisco had anything to do with knowing about her full-blown affair with Larkyns, though he certainly disliked the man having any contact with her at the time. Instead, he claimed he had sent Flora to Oregon so she could have help with young Floredo in his absence.

Muybridge told his wife he would allow her $50 a month while in Oregon, which he reckoned was plenty to support her and the baby comfortably, though he also showed a surprisingly un-Victorian attitude to a young mother working: "If she wanted more she could easily

earn as much as that by her wax-work, at which she was very proficient, and by touching up photographs which she did pretty well."

Soon after her departure Muybridge had a series of meetings with
a midwife named Susan Smith (interestingly reminiscent of
Muybridge's mother's maiden name, Susannah Smith). These meetings were the trigger that initiated the attack on Harry Larkyns. The
first encounter was a formal, acid-and-ice affair in a lawyer's office in
the expensive first district of downtown San Francisco. Muybridge had
lost an imprudent fight, one he should never have picked.

At the heart of their petty legal battle was Smith's occupation. It
was Susan Smith who had seen Flora Muybridge through her pregnancy from the early stages, and it was she who had delivered the
Muybridges' baby son Floredo Helios Muybridge in April of 1874.
Smith had appeared to carry out her duties for the Muybridge family
with enthusiasm, even with affection.

Flora needed the help. During the pregnancy, Eadweard was away
for weeks at a time on photographic expeditions that took him into the
farthest reaches of the western wilderness. In his absence Flora was
reliant on Susan Smith. The midwife had been by Flora's side when she
unexpectedly went into labor before Eadweard had returned from his
latest expedition. It was Smith's hands that had eased the baby into life,
cut the cord, and wrapped Floredo in a blanket for his mother to hold.
She had been there for a good 24 hours before Eadweard Muybridge
could get back to town to see his newborn son. It would seem that
Muybridge owed Smith a large debt of gratitude. Yet once the baby was
delivered, he refused to pay her bill.

What possessed a man like Eadweard Muybridge to default on a
$100 debt? He was not short of cash; the Muybridges were comfortably
well off by the standards of the day. Nor was this a matter of simple
forgetfulness. Muybridge could be eccentric, but he managed the fi-
nances of his household and his business with true Victorian rigor.

Equally, Susan Smith was not incompetent, though she had little
formal training. In 1870s America, midwives were yet to become medical professionals. They were direct descendents of medieval
wisewomen, reliant on experience that was passed down from generation to generation. Where their European counterparts were already

by this time specialist nurses, U.S. midwifes like Smith were limited to traditional cures and herbal remedies. She could offer practical experience and a helping hand for the mother, but no medical guidance.

The usual clients of midwifes were the urban poor, or families living too far from a city to get clinical help. Middle-class parents like the Muybridges would normally bring in a physician to see them through the birth. Yet it was Susan Smith that had been chosen to help with Flora's pregnancy. Perhaps it was Eadweard's European expectation of a midwife's professionalism that led them to take her on, or perhaps it was pressure from Flora, who was a good friend of Smith's daughter, Sarah Louisa. Despite this family friendship though, and despite Smith's monotonously regular requests for payment, for six months no money was forthcoming.

Eadweard claimed in court that he had not settled Smith's bill because he had already given his wife the money to pay the midwife and considered the matter closed, but there was more to it than this. Smith had a sly, conspiratorial manner, several times hinting that she knew more than she was telling. And Muybridge had seen her whispering in corners with Harry Larkyns—a man he suspected of having designs on Flora.

The dispute between the obstinate Eadweard and the devious Susan came to a head at the start of October. Smith would wait no longer to be paid; she resorted to the law. With the help of her attorney, she took her case to the Justices' Court. Eadweard could only argue that he considered the bill already to be paid. After a brief hearing, the judge ruled that Muybridge should pay Smith a total of $107 or face prison for contempt.

During this court session, Smith had produced a letter from Flora to prove the truth of her demands. In it, according to Smith, Flora had mentioned Larkyns with some affection. After the hearing, it was the letter that stuck in Muybridge's mind, rather than the true purpose of the lawsuit. Could Flora have betrayed him? Had she some guilty secret that Smith was covering up?

On the following morning—October 15th, forty-eight hours before the fateful day when Eadweard would set off for the fatal meeting at the Yellow Jacket ranch—a meeting was arranged to settle costs at the office of Muybridge's attorney, a Mr. Sawyer. Muybridge and Smith met in the street outside the lawyer's office. Smith later recalled their meeting:

> I saw him first in the morning in front of Sawyer's office, pacing up and down and looking wild and excited. I met him by appointment and said: "I have come to see about getting my money." He wanted to know if I had any proof of his wife's guilt. I replied, "I don't know." He said: "Mrs. Smith, if you don't tell me the truth I shall consider you a bad woman."

Together, stiffly, Muybridge and Smith went up the stairs to his attorney's office. They sat across a table, and Muybridge handed a small package across the dark, polished wood. In it was the $107 in cash. The debt was settled. Almost as if in return for the payment, Smith passed over two letters, one from Larkyns to Flora Muybridge, and another from Flora to Sarah Smith, the midwife's daughter.

Muybridge did not read the letters, which were handed to Mr. Sawyer. Instead he waited until the following evening, October 16th, and paid a call on Susan Smith. The midwife opened the door herself and led Muybridge into her parlor. He refused the seat Smith offered him. Standing tall over the woman, he demanded to know the truth. According to Mr. Sawyer, he said, the letters did not demonstrate anything more than a mild flirtation on his wife's part. Was there anything more? Their conversation can be reconstructed from accounts that were later made of it in court. Initially, Smith did not answer. Muybridge's temper was rising.

"What were you doing with Major Larkyns, the day I saw you talking to him at Clay and Montgomery? Didn't you carry a note from Larkyns to my wife?"

Smith nodded. Her "yes" was practically inaudible.

Now the truth was coming out. And it had to be about Major Harry Larkyns. Who else but Larkyns? Larkyns, the "friend" who had entertained Flora when the photographer was on his regular expeditions out of town. Major George Harry Larkyns, a popular man-about-town but with a dangerous reputation. A man Muybridge had already warned not to spend any more time with his wife.

Standing silently in the doorway of her parlor, Muybridge stared at Susan Smith as if she were something less than human. "Why did you say nothing? Why did you not tell me the truth of it that evening?"

"When you asked me what I was doing with Major Larkyns, Flora was standing behind your back. She was so pale. She shook her finger warningly at me. I supposed something was wrong, so I told you nothing."

Muybridge's face drained of all color. He stamped on the floor like a frustrated child and shook all over. "Mrs. Smith, you know more than you tell me. How could the woman I loved so dearly treat me so cruelly?" Muybridge was beside himself. Smith later said that he talked incoherently for a while after this outburst and then went away.

He returned the next day, on the late morning of October 17th, just before noon to wait on that doorstep a final time. Smith came to the door. She saw a man she described as looking "even more terrible than I had ever seen him. He appeared as though he had had no sleep the night before."

It's not surprising. The day before, Muybridge had not opened the two letters Smith handed over to his lawyer. But in the intervening hours he had brought himself to read them. These were no simple indications of a "mild flirtation" as the attorney had suggested. The first was from Harry Larkyns.

Dear Mrs. Smith,

You'll be surprised to hear from me so soon after I left the city, but I have been so uneasy and worried about that poor girl that I cannot rest, and it is a relief to talk or write about her. I think you have a sufficiently friendly feeling towards me to grant me a favor, rest assured I will find means to do you a good turn before long. I want you and your girls to be perfectly frank, open and honest with me. If you hear anything of that little lady, no matter what, tell me outright. She may return to the city and beg you not to let me know, but do pray not listen to her. Do not be afraid that I shall get angry with her, I will never say a harsh word to her, and even if things turn out as badly as possible and I find that she has been deceiving me all along, I can only be very grieved and sorry, but I can never be angry with her.

Telling Smith that he has left Flora messages in the personal columns of newspapers, most notably "Flora and Georgie [Floredo]: if you have a heart will you write to H. Have you forgotten that April night when we were both so pale?" Larkyns went on:

Mrs. Smith I assure you I am sick and ill with anxiety and doubt, the whole thing is so incomprehensible and I am so helpless. I fear my business will not let me go to Portland and I see no other way of hearing of her. If an angel had come and told me she was false to *me*, I would not have believed it. I cannot attend to my work, I cannot sleep, and the longer matters stay like this the more I suffer, besides even if she does write now, I shall not know what to believe. I cannot help thinking of that speech of hers to you the day before she left, when she begged you not to think ill of her, whatever you might hear, it almost looks as if she had already settled some plan in her head that she knew you would disapprove of. And yet Mrs. Smith, after all that has come and gone, *could* she be so utterly untrue to me, so horribly false? It seems impossible, yet I wrack my brain to try and find some excuse, and cannot do it. If she had nothing to conceal why has she not written? If I got to Portland I must give up my situation, but I think unless we hear I shall go.

Larkyns seemed genuinely devastated that Flora had cut off all communication. This was no mild flirtation. Flora's letter to her friend Sarah, written on July 11th, was equally damning.

Dear Sarah,

Yours of the 3rd has just come to my hand. I had begun to think you had all forgotten me, I write again to your Mother. I received such a letter from H. L. saying that he heard I had not gone any further than Portland. . . .

Flora was staying with her uncle's family in The Dalles at the end of the Oregon Trail. It seems Larkyns thought she had stopped off en route in Portland, and Flora was worried that he would write letters to her there, which might fall into the wrong hands. Later in her letter she wrote:

I am not ashamed to say I love [Harry] better than anyone else upon this earth, and no one can change my mind, unless with his own lips he tells me that he does not care for me any more. . . .

She did not want him to come to her though, as "this is a small place and people cannot hold their tongues." Flora ends by entreating Sarah to "destroy my letters after reading them, for you might lose one and it might get picked up."

What Susan Smith had done was hardly innocent. She had deliberately ensured that Muybridge knew what had happened between his wife and Larkyns. Still, the social niceties were observed. On his arrival after reading the letters, Muybridge asked, "Mrs. Smith, are you busy? I want to see you."

Smith told him she was not busy and asked him in. Was it the light of day that showed him something in her house that he had not seen the evening before? Or had Smith brought something out that she had kept hidden? Either way, as he came into her parlor he crossed straight to a table in the middle of the room and picked up a photograph. In his face was a real shock of recognition. Again the conversation is reconstructed from trial accounts.

"Who is this?" Muybridge asked, though he knew very well.

"It is your baby," said Smith.

Muybridge seemed confused. "I have never seen this picture before. Where did you get it? Where was it taken?"

"Your wife sent it to me from Oregon," said Smith. "It was taken at Rulofson's."

There was something wrong. He was the state's leading photographer. Why did he know nothing of this portrait? Muybridge fingered it carefully, as if it were a strange, potentially dangerous artifact. He examined it closely, taking in every detail. The eyes, the wisps of hair, those tiny fingers. Floredo, his own son. He still felt a frisson of excitement at the thought that he, he and Flora, were responsible for bringing this tiny human being into existence.

Just a few years before he would have found it difficult to imagine ever having a child of his own, but then Flora Shalcross Stone had entered his life. When she came to the studio for a job retouching photographs, Muybridge had fallen instantly, unquestioningly in love with this resolute young woman. Her youthful vibrancy was a good match for his passion for his art; their 21-year age difference seemed not to matter to either of them. Now, despite all their problems, they had Floredo to unite them.

Then he turned the picture over. His face, behind the beard, went very pale and then reddened.

"My God! What's this?" Clearly written on the brown backing paper of the picture, in his wife's rounded handwriting, were two words.

Little Harry.

Smith watched aghast at Muybridge's reaction to the picture. She might have assumed that there would be some sense of triumph in bringing her ex-employer down a peg, but now what she saw frightened her. She would tell in her later court appearance:

He stamped on the floor and exhibited the wildest excitement. His appearance was that of a madman; he was haggard and pale; his eyes were glassy; his lower jaw hung down, showed his teeth; he trembled from head to foot and gasped for breath; he was terrible to look at.

Muybridge had no self-control left. He cried out, "Great God! Tell me all!" and came toward her like an automaton with his hand raised to strike. Now the fear flooded through Smith.

I said "I will tell you all." I thought he was insane, that he would kill me or himself if I did not. I then told him all I knew.

She had plenty to tell. She spoke quickly, as if trying to get the words out before Muybridge could attack her, but now he stood frozen before the onslaught of the truth.

"Mr. Muybridge, your baby was born, as you know, on April 15, 1874. Before its birth, Mrs. Muybridge was afraid it would have sandy hair, like Larkyns. After it was born, she asked me if it had sandy hair. I replied, 'Its hair is light.'

"Just before its birth, the same evening, Larkyns and Mrs. Muybridge came to my house in a carriage together. He called me out and made me go along in the carriage with them to Mrs. Muybridge's home. You were not, as you know, at home, Mr. Muybridge. Larkyns held her in his arms and kissed and caressed her. Then Larkyns went for a doctor and brought him to her house. He then went out again."

Muybridge stood as still as stone throughout the revelation. But the redness of his anger drained again from his face until it seemed entirely devoid of blood.

"After a while, Larkyns came back to the house, went to Mrs. Muybridge's bedroom, stooped over the bed, kissed Mrs. Muybridge, and said, 'Never mind baby, it will soon be over.' He meant Mrs. Muybridge by 'baby.' He then went away."

This wasn't the end for the anguished Muybridge. She continued to pile on the painful details.

"Three or four days later he was back. Mrs. Muybridge ordered me to bring the baby for Larkyns to see. She asked him, 'Who do you think the baby's like?' Larkyns smiled and said, 'You ought to know, Flo.' She laughed and made no answer.

"A few days later Mr. Larkyns was at the house again and said to me after he came out of Mrs. M's room, 'Mrs. Smith, I want you to take good care of that baby. I hold you responsible, for I have two babies now.' I laughed, and when he'd gone I told Mrs. Muybridge I didn't know what he meant.

"On one of his visits, Larkyns was standing at her bedside, and she said 'Harry, we will remember the thirteenth of July. We have something to show for it.' She and Larkyns looked at the baby and smiled. Larkyns used frequently to be in her bedroom and stay there for hours. And they wrote to each other two or three times a day."

She told Muybridge how Larkyns and Flora had practically ignored her, treating the midwife as part of the furniture. She had, she said, often been present when Larkyns called. Muybridge heard how Harry had not confined his attentions to outings to the theatre. He was a frequent visitor at the Muybridge home. In fact, when Eadweard was away on his trips, Harry seemed always to be calling. At other times, Flora had ventured to the major's apartment unchaperoned. Although Smith did not know it, it would later turn out that this flat was rented in the name of Mr. and Mrs. Larkyns—yet there was no "Mrs. Larkyns." Flora had been living with Harry as man and wife.

After Smith's relentless tirade of exposure, it was the mundane, everyday nature of her final tidbit of information that brought the truth home to Eadweard. He asked Smith if she could tell him what his wife did with the money he gave her, the money that should have been used to pay Smith's bill, and she replied: "It may be that the extra washing that was put in accounts for it."

Muybridge was bemused—this seemingly irrelevant comment surely had nothing to do with the situation. "What do you mean?"

"Major Larkyns sent his old shirts to your house to be repaired and done up. They were sent to the wash with yours, and the man who carried letters between him and Mrs. Muybridge came and took his shirts to him after they were washed. And your wife paid for the washing."

Muybridge, his face still pale as ice, was now shaking uncontrollably and collapsed to the floor. His fall was unbroken; there was a loud crash as he hit the floorboards, and then he lay still, so still that for a moment Smith thought that he had died.

She rushed for her smelling salts. Muybridge's head jerked up, but for a while he remained on the floor. He was not Floredo's father. The sensation was crushing. It was like a bereavement. He had entered the house a father—now he had been robbed of his son. He had lost his child and his wife, too. And this was all the fault of the despicable Major George Harry Larkyns.

Climbing to his feet, in a voice that had icy control, Muybridge asked Stone if she knew where Larkyns now was. The last she had heard, it seems, was that he had traveled out to Pine Flat, beyond Calistoga, around 60 miles away as the crow flies, at the far end of the Napa Valley. Eadweard was expressionless as he took in the information, but lost some of his calm as he reached the door.

Smith would later tell the court:

> He muttered to himself and then I spoke to him again. He turned as though he had awakened from a trance. He said: "Flora, Flora, my heart is broken. I would have given my heart's best blood for you. How could you treat me so cruelly?" He trembled like a leaf in every limb. His under jaw hung down and he showed his teeth. His excitement was intense. I thought, and my opinion is, that when he left me he was insane.

Ↄ☉ↄↄ

The walk back from the Smith house into the center of the city gave Muybridge a chance to regain his self-control. Just before one in the afternoon, as he neared the Bradley and Rulofson gallery, Muybridge ran into an acquaintance, the actor Harry Edwards. Edwards saw nothing unusual in Muybridge, even thought him cheerful. But this carefully constructed pose did not last long when he found his friend William Rulofson inside the gallery. Muybridge had already been upstairs, and Rulofson first saw him riding down in the elevator.

This apparently trivial occurrence was enough to make Rulofson concerned. Muybridge had a profound dislike of the elevator. On a number of occasions in the past, when Rolufson and he had walked into the building, Muybridge had left his friend to travel in the elevator while he ran up the stairs. It seemed particularly strange to Rulofson that Muybridge, who would not even make use of the elevator on the way up, was now making the easier journey down in it.

"What is it?" Rulofson asked him. "What's the matter?"

"You will find out soon enough," Muybridge replied.

He was clearly distraught. Worried about a possible scene in front of his customers, Rulofson bustled Muybridge into his office. By now, as Rulofson would later describe in court, tears and perspiration were streaming down Muybridge's face. Without any explanation, Muybridge launched into a strange plea. "Do not ask me to name my dishonor. Promise me, Rulofson, that in the event of my death you will faithfully give my wife all that belongs to me."

He seemed to Rulofson to be on the verge of suicide. The photographer tried to persuade Muybridge to change his mind, to make him understand just how good things were for him at the moment—with a new, young family, success from his photographs, the world at his feet.

Muybridge said that he had no thought of "such a cowardly intention" as suicide, but wanted to vindicate his honor even if he might lose his life in doing so. He thrust two letters, the letters his lawyer had given to him from Smith, into Rulofson's hand, asking him to keep them safe until he came back, or to burn them if he should die. He told Rulofson that in the letters was absolute proof that he had been dishonored by his wife. Because of this he had to see Larkyns, and Larkyns was at Pine Flat. The boat for Vallejo, on the way to the Napa Valley, left at 4 p.m.

As Rulofson tried to make sense of what he had heard, Muybridge left for his own office on the far side of the building. Rulofson kept watch on the office, leaving Muybridge alone, hoping he would see sense. When, around 2:30, Muybridge came to his door, Rulofson hurried over, hoping to delay Muybridge long enough so that he would miss the ferry. They would talk for over an hour. As the time ticked on toward 4 p.m., Rulofson tried to defend Flora.

"Many good women are basely slandered."

Muybridge ignored him. He seemed to Rulofson to have withdrawn from reality. "One of us will be shot," he announced with unnerving calm, showing Rulofson his pistol.

"For God's sake, don't kill him!" Rulofson shouted in his face. Muybridge put the pistol away, evaded Rulofson's attempt to hold him back, and started on his run down to the waterfront.

The train of events that began with the failure to pay Susan Smith would bring Muybridge to Harry Larkyns's murder and to the Napa jail, to await his trial for murder.

FIVE

TRIAL

This all absorbing object of public interest came on in the regular course of business for trial today. The Court room, as might have been expected, was well-filled with spectators, though so far as the preliminary proceedings have yet gone, it was not crowded. As far as we shall be able to report in today's REGISTER nothing pertaining to the substantial progress for the case has yet been done. . . .

The defense were ready, but the council for the people, Dennis Spencer Esq., District Attorney, announced the unavoidable absence from sickness of one of his witnesses, and the unaccountable delay of others and asked a continuance until they could be procured. The parties absent were George Wolfe, driver for Foss & Connolly, who drove Mr. Muybridge up to the mine from Calistoga on the night of the tragedy, and J. M. McArthur, a prospector, who was at Stewart's house at the Yellow Jacket when the shooting occurred. . . .

Wolfe has been down with the measles and is not yet able to be out. McArthur is now engaged at the Empire Mine, Lake County, and his absence was not accounted for by anyone.

Napa Daily Register, February 2, 1875

After the terrible events of October 17, 1874, despite Muybridge being locked away in the jailhouse, there were still loose ends to be sorted out. What was to become of Flora and Floredo? And, most immediately, there was the funeral of Harry Larkyns. Not surprisingly, the newspapers picked up the story with glee, not just because an eminent man like Muybridge was involved but also to revel in the less-than-ordinary life of Harry Larkyns. And it soon became obvious that even the story we have already heard was less than complete in putting together a full picture of this charming if occasionally dangerous rogue.

Take, for instance, his job at the *San Francisco Post*, the position of theatre critic that he had held when he was first introduced to the Muybridge family. While it's true he occupied the post, it turned out he got it—and lost it—in bizarre circumstances.

According to a report in the *San Francisco Examiner*, Larkyns had met and taken pity on a man called Coppinger he found "shivering and hungry in the streets." Although Larkyns was dishonest, he seems also to have been a generous man, as he took Coppinger in to share his rooms. In exchange, though, Larkyns got Coppinger to write his theater reviews for him. Larkyns seems to have been very happy to sit back and trust his reputation to the writing skills of a man he picked up off the street, emphasizing an approach to life that sat somewhere on the scale that runs from happy-go-lucky through complacent to downright stupid.

Word reached the *Post* that Larkyns was not writing his own pieces, an allegation he didn't try to deny. It seems likely that Coppinger, who got Larkyns's job after he was fired, had written the anonymous note that betrayed his friend. It seems certain, from the way he treated Coppinger, that Larkyns believed this. As the *Examiner* tells us:

> From that time on, whenever Coppinger met Larkyns, Larkyns would take him by the nose and the lower jaw, and spreading open his mouth with a grip of iron, would spit down Coppinger's throat! He was arrested for this and tried. When the jury heard the story of Coppinger's business, they acquitted Larkyns amid the cheers of the spectators.

Remember that though Larkyns had helped Coppinger off the street, he was still defrauding the newspaper by getting Coppinger to do his job for him, but Larykns's charm managed to keep his deceit from the mind of the jurors.

When news of Larkyns's death reached San Francisco it soon came to the ears of Coppinger, who by now had been given the nickname Cuspidor by the newspapers. Although it is entirely apocryphal, the story of Coppinger's response to the news made it into the *San Francisco Examiner* as fact. Pulling out all the stops, the reporter tells us that Coppinger pushed aside his glass of beer and ordered something stronger.

"There is, indeed, a special Providence in the killing of that man. I drink to the special Providence," he is reported as saying. This toast so disgusted his fellow drinkers that those around him put down their glasses and walked away from the bar. But Coppinger hadn't finished. He hurried over to the undertakers and "gazed long and maliciously at the features of the dead." When he had finished gloating, Coppinger rubbed his hands, grinned like a jackal, and announced to the world that he would walk 20 miles on the stormiest night to see Larkyns in his coffin.

It was probably only out of consideration of how much the public would stand that prevented the newspaper from accusing Coppinger of using his old enemy's corpse as a spittoon. As far as the press was concerned, Larkyns was a popular figure, almost one of their own, and to be protected. Overlooking the fact that Larkyns himself was hard drinking, the writer was deliberately casting Coppinger as a degenerate drunk.

Some of the obituaries Larkyns received seemed to be written about a totally different man from the conman and rogue that he had been in life. These effusive extracts from a lengthy piece in the *San Francisco Examiner* are typical:

> Larkyns was over six feet tall, straight as a lance, and had the gift of spreading a ripple of sunshine wherever he went. His wit and clever stories and general affability won him a legion of friends.
>
> He was every inch a bohemian and debonair man of the world. He spoke divers tongues, all equally well as he did English. He had been everywhere and seen everything. He had roamed the world, and been a soldier of fortune, writer, poet, musician, the Lord only knows what he had not been.

It's just possible that the writer's tongue was in his cheek, describing a man with Walter Mitty-like fantasies, but all the evidence is that these achievements, largely only ever evidenced in Larkyns's stories, were taken as the gospel truth. The gushing, up to now only moderate in force, was then turned up to full strength:

> He could box like Jim Mace and fence like Agramente, and he could outfoot them all in a race. He could hit more bottle necks with a pistol at twenty paces than anyone else, and he never sent his right or left into a bully's face, but that the bully was carried away on a shutter.

Larkyns was a scientist, chemist and metallurgist lecturer. No one could mention anything he could not do better than anybody else, and when it came to cooking a delicacy in a chafing dish, Dalmonico was simply not in it. A sniff and a shrug of the shoulders from him would put any brand of wine out of commission with the epicures of the Comstock.

On another occasion he sat down at the cathedral organ and improvised such a melody that the crowd, rather than walking out to his music, stayed, enraptured, to listen. Has there ever been such a paragon? Yet there is nothing in the article that suggests irony. The obituary goes on to give a wonderfully garbled version of the events leading up to the murder—rather than the midwife Smith being owed $100 for her services we hear that a letter between Larkyns and Flora Muybridge "fell into the hands of a maid in the Muybridge home, she sold it to Muybridge for $100."

It was in this obituary also that Muybridge's apology to the ladies in the Yellow Jacket ranch house was handed over to Larkyns, giving his departure from this world the sort of romanticism that would later be a trait of the worst excesses of Hollywood:

> Larkyns staggered back into the room and with his hand covering his heart, which had been pierced by the bullet, bowed low to the ladies. "I am very sorry this little trouble has occurred in your presence," he said with the hesitation of death in his tones, but with a soft smile playing about his lips, and then, straightening up, started for the door saying, "Kindly excuse— me—for—a—moment." With this he passed through the door and reeling, fell dead.

With a soft smile playing on his lips? Even Mills and Boon wouldn't stoop so low. Larkyns's body was held for a short while at Lockhart and Porter's, an undertaking firm based at 39 Third Street. His friends arranged for an elegant rosewood coffin with silver (or at least silver plated) handles and screws. Those who came to view the body observed that he was still dressed in the business suit in which he'd left San Francisco for Calistoga. His face was bruised and his shirt bloody, with a clear bullet hole accurately marking the position of his heart.

The major was given a respectful send-off by the Reverend Henry D. Lathrop in the Episcopal Church of the Advent in San Francisco on October 19th, only two days after his death, and before the story behind the murder had seeped out. Even now, Larkyns continued to exert an influence over the ladies—according to the *San Francisco Chronicle*, a "well-known actress" from the California stage jumped out of her seat and ran out, sobbing, to lay a wreath on his coffin as it passed down the aisle. The choir accompanied his entrance with his favorite hymn, the largely (and probably sensibly) forgotten saccharine *Flee as a Bird to the Mountains*, accompanying the "great Baritone" George Russell.

The oration, ironically given by the same Harry Edwards who had chatted to Muybridge as he set off to kill Larkyns, continued the effusiveness of his obituary, praising the dead man for being "a gentleman in the truest sense of the word," a man for whom "vulgarity of every kind was a perfect stranger to his soul." A very different Larkyns to the truth that would emerge at Muybridge's trial. From the church his remains were taken to the Masonic cemetery.

Muybridge was to sit in jail for nearly two further months before being indicted for murder by a grand jury on December 8th. His indictment, signed by District Attorney Dennis Spencer, read:

> The People of the State of California against Edward [*sic*] J Muybridge in the County Court of Napa County, State of California. December term A.D. one thousand eight hundred and seventy-four. Edward J. Muybridge is accused by the Grand Jury of the County of Napa, State of California, by this indictment of the crime of Murder.

> Committed as follows to-wit: The said Edward J. Muybridge on the 17th day of October A.D. 1874 and before the finding and presentation of this Indictment did, at the Country of Napa and State aforesaid feloniously willfully and unlawfully and of his malice aforethought kill and murder one Harry Larkyns contrary to the form of the Statute in such cases and made and provided, and against the peace and dignity of the People of State of California.

He only spoke to say "not guilty." His counsel, the highly experienced Cameron King of San Francisco and the gifted Wirt Pendegast of Napa, entered a plea of insanity. It seems likely that Pendegast (and perhaps King) offered to take on the role unpaid—Muybridge would later write of his defense lawyer's "noble and disinterested generosity."

One week later, Flora Muybridge made one of her last ventures into the limelight. She went to court in an attempt to get a divorce from Muybridge, claiming extreme cruelty. Flora's idea of cruelty was the fact that Muybridge had "looked at her when she was in bed to make sure that she was there." Despite two attempts, the court would not award her any alimony based on this story. As far as we know, Muybridge was never to have any further contact with her.

For the most part Muybridge remained silent, speaking only to his legal team. The only break in this self-imposed isolation came around the same time as Flora's attempt to divorce him. He spoke to a reporter, concerned it seemed to preserve the good name of his business even under these difficult circumstances. He had read in a newspaper that he was employed by Bradley and Rulofson, and wanted to make it clear that he was not an employee, but was running his own business based at their premises.

Muybridge's intention may have been merely to clarify his social position, but it is likely that the reporter, George Smith of the *San Francisco Chronicle*, had hopes for rather more. He had turned up in Napa to see Muybridge, a long shot as the prisoner had already announced that he had no interest in speaking to reporters.

Smith turned out a report that was largely favorable to Muybridge, despite the decidedly imbalanced headline, "The Fatal Amour: an Interview with Muybridge, The Slayer of Harry Larkyns." After giving a little background to the incident, he opened by describing a man far removed from the image of a violent killer:

> Confinement and care had made him paler than usual, but he appeared to be in good health and excellent spirits. Muybridge is forty years of age, but looks at least ten years older than that. His full, unkempt beard is deeply tinged with gray, and his hair is white. He has mild blue eyes and a face which a physiognomist would invariably pass by in searching for one likely to do deeds of violence or death. His manner is quiet and reserved, his dress plain to a degree somewhat out of keeping with his profession and standing, and any one unacquainted with him would readily mistake him for a quiet, good natured old farmer.

Smith then asked Muybridge questions about suggestions that Muybridge was suspicious of infidelity when he sent Flora to Oregon, and that he knew her character to be doubtful before he married her and had encouraged her to become estranged from her husband. Muybridge acquitted himself well in his replies. He was careful not to criticize Flora more than he must. He said that she told him that the trouble between her and Stone [her first husband] was due to inequality of age (Flora was still only 23 at the time of the interview).

Flora, said Muybridge, had been driven from her first husband by incompatibility of temper and the cruel treatment afforded her by Stone and his mother. He had never heard any criticism of her chastity before this affair. One of the few times in the interview he would become emotional was when he now declared that he had loved Flora with all his heart and soul, and that discovering her infidelity was a cruel, prostrating blow. "I have no fear of the results of the trial," he continued. "I feel that I was justified in what I did, and that all right-minded people will justify my action."

It does not seem to have occurred to Muybridge that the substantial difference in age between him and Flora could have been a problem. This remark that "all right-minded people will justify my action" was picked up by another paper, the *Sacramento Bee*, which commented with the literary equivalent of severely raised eyebrows that this statement amounted to "a severe rebuke to those who did not believe in murdering those who offend them."

In the interview with Smith, Muybridge also described the way he went after Larkyns, half expecting to be killed himself. He said that he took the first boat to Vallejo intending to punish Larkyns. He had no idea whether he would come back from Calistoga alive. He expected to find Larkyns armed. As Muybridge quaintly remarked, "I supposed that such men would always be prepared for the consequences of their wrongdoings." Muybridge didn't expect a single shooting, but rather a fight, perhaps with several others and that he, himself, would probably also be shot, even if Larkyns were killed.

He was quite open about the moment he killed Larkyns:

> I saw that he would be gone in a moment and that I must act on the instant or he would escape me; so I fired on him. I did not intend to shoot him at once, but thought to parley with him and hear what he had to say in excuse or extenuation, but he turned and ran like a guilty craven when I pronounced my name and said I had heard about my wife and I had to shoot him or let him go unpunished. The only thing I am sorry for in connection with the affair is that he died so quickly. I would have wished that he could have lived long enough at least to acknowledge the wrong he had done me, that his punishment was deserved and that my act was justifiable defense of my marital rights.

Muybridge had nearly two more months to think about his defense, justifiable or otherwise, before the trial opened in the Napa courthouse on February 3, 1875. On his side of the court was still the team of King and Pendegast. Against them, prosecuting, were not only the district attorney, Dennis Spencer, but also Judge Thomas P. Stoney, who was assisting Spencer because he claimed limited experience in this type of case. On the bench was William T. Wallace, at the time the chief justice of the Supreme Court of California.

The trial opened with Muybridge's plea of not guilty. According to one newspaper report, he turned to the sheriff standing by him and commented with a quiet laugh, "To kill a man and yet plead not guilty!" The jury, all men, mostly grizzled old-time westerners, were then chosen. According to the report, it seemed at first as if this could take some time as three were "excused for incompetency" and 14 "peremptorily challenged," but despite this the impaneling was "accomplished in a remarkably short time, considering the importance of the case."

Tellingly, Judge Stoney, despite technically being the assistant, opened for the prosecution with a powerful speech establishing that self-defense was no excuse under the law—and defending your wife's honor was, in effect, self-defense—and that the Muybridge case was a clear-cut example of murder. Suspecting that the jurors might be sympathetic to the defendant, he made it clear that this was not so much a comment on the rights and wrongs of what had been done but an assessment of the truth. The *Napa Daily Register* reported his opening statement to the jury:

> . . . one of the cleverest, most earnest and eloquent appeal [*sic*] to the right feelings of men's natures it has ever been our good fortune to listen to. He said there had been an old formula of English law he could wish had been retained. He said when a prisoner was arraigned before a bar of justice he was saluted with the greeting "May God grant you a good deliverance." But though the form was changed the spirit remained the same. No personal feelings are cherished in the prosecution. No vengeance, but a simple desire to vindicate the majesty of the law, and judge strictly of the prisoner's acts according to the light of the statute. He enjoined the jury to be inflamed by no outside considerations but take solely the fact of the killing and inquire whether the law justified it.

Like all the other quotes on the trial below, this one is from a contemporary newspaper report, which almost certainly abridged the ac-

tual words used. The prosecution witnesses clearly and simply established that Muybridge had committed the act as a cold-blooded, calcu-
lated attempt to kill Larkyns, something that after all was not difficult,
given that Muybridge had never denied that this is what had happened.
First up was the medical examiner, S. J. Reid. His testimony was practical and to the point. He found Larkyns dead, and examined the body.

> The orifice was $1^1/_4$ inches to the right and below the left nipple and about
> 1 inch below the sternum. The wound was on the left side, ranging inwards
> and upwards. The ball had penetrated the heart, causing hemorrhage which
> evidently caused death. The wound was necessarily a fatal one. No man
> could live over 30 or 40 seconds after receiving it.

There was no cross examination; the defense had nothing to deny.
Muybridge admitted shooting Larkyns with the fatal bullet. Next on
the stand was A. H. Connolly, the joint owner of the Foss & Connolly
livery stables in Calistoga where Muybridge had hired the buggy and
driver to take him out to the ranch. According to his testimony,
Muybridge arrived at the stable on the night in question soon after the
train pulled into town, wanting a horse to go to Pine Flat. According to
the *Napa Daily Register*, Connolly told Muybridge it was a bad time to
go there.

The night was dark and the road muddy. Connolly urged
Muybridge to wait until morning for the stage coach that routinely
rattled up to Pine Flat. But Muybridge was adamant. By chance, he
mentioned that his business was with Harry Larkyns. Connolly raised
an eyebrow—he had seen Larkyns in Calistoga only the day before. He
turned to William Stewart, the superintendent of the Yellow Jacket
mine, who was picking up his horse after a trip into town. Stewart
confirmed that Larkyns wasn't at Pine Flat. In fact he was staying at
Stewart's own house, the Yellow Jacket ranch.

Stewart told Muybridge that Larkyns would be back in Calistoga
the next day if he cared to wait, but Muybridge insisted on leaving at
once. Connolly sent for a young man, George Wolfe, to drive him,
rigged a team, and sent them off into the night. The witness had placed
Muybridge on his way to the murder scene. The picture at the ranch
was given by B. S. Prickett, Stewart's foreman at the Yellow Jacket ranch
(with typical inaccuracy, the newspaper refered to the ranch owner as
Stuart, and Larkyns as Larkyn):

He was at the ranch on the night of 17th October. Two of these men, Murray and McCrory, had been to the Missouri Mine for quicksilver, and they two and witness had just stepped out to see how much they had got. Were standing in the back door when a team drove up to within about 50 feet of the door, and a man got out, advanced to near the door and asked for Larkyn. Witness asked the stranger in, but he declined, saying he would not detain him but a moment. Some one had spoken to Larkyn and he came out.

The reporter went on in the same terse style to describe Larkyns peering out into the darkness to where the voice had been heard. According to Prickett, Larkyns said, "It is so dark I cannot see you."

Prickett went on to describe Muybridge's ironic salute: "My name is Muybridge, and I have a message for you from my wife." Now came the first mention of Larkyns's alleged career through the house, with his hand clapped over his heart. According to Prickett he ran from the back door, through the hall to the front door with Muybridge in pursuit. When Prickett next saw the prisoner he was seated in the lounge, under arrest with the gun of one of Prickett's colleagues pointed at him.

Prickett's evidence might not seem too inconsistent with the medical examiner's story that "no man could live over 30 or 40 seconds" after receiving this injury—time to charge through the house and collapse against the tree—but medical assessment since has suggested that even if Larkyns had lived that long he would not have been able to run (or walk) anywhere. Looking at a photograph of the ranch house, which no longer stands on the site, the oak tree appears to be by the back door, not the front—and the back door is the first that would be approached from the drive, so it seems more likely that he was shot by the back door and collapsed against the tree where he stood.

The defense attempted to query some details of the evidence, but with little material effect in Prickett's response. With a supporting description of events from another witness, James Murray, the court recessed until the next day, Thursday.

This second day of the trial started with a bit of business that verged on farce. The *Napa Daily Register* takes up the story:

Precisely at 9:30 Judge Wallace took his seat, rapped on the desk and called the court to order. "Court's in session," said Under-Sheriff Palmer. The Clerk called the jury roll and found all present. Here proceedings were interrupted by a party who wanted to be naturalized, which was briefly attended to and another citizen added to the support of the Administration. Muybridge appeared, looking as well as usual, and pending the little naturalization by-play, counsel stroked their beards and called up tender memories of the breakfast just discussed.

With the formal, precise proceedings of the modern court system it can be hard to remember that they used to mean it literally when they asked "anyone with business with the court" to attend, and individuals had the liberty to interrupt even a trial of this magnitude to make a request of the judge. That this was seen as a big trial is obvious from the description of the opening of the second day in the *San Francisco Chronicle*. The courtroom, it seems, was crowded with spectators all day, and a good number who couldn't get into the room hung around the door and crowded the hallway. The paper remarks that some of those present had come especially by train to witness the spectacle.

Once the naturalization business was out of the way and the case was back on track, C. A. McCrory was called as the other witness of the shooting and came up with a subtle variant on the story:

> Saw the difficulty between Muybridge and Larkyn. It was on the 17th of October, about 11 o'clock, a gentleman came to the door and asked if Larkyn was in; I said "yes." Think we all three answered "yes" about the same time. Some one spoke to the Major, who came to the door and spoke to the stranger. I hardly know what was said except that Larkyn said that it was so dark that he could not see him. After that I heard the pistol shot and the Major ran through the door.

McCrory described McArthur stopping Muybridge as he entered the parlor. McArthur then asked McCrory to check on Larkyns: "See where the Major is; see how bad he is hurt." Murray, another of those present, wanted to carry the fallen Larkyns back into the parlor, but McCrory claimed he replied, "Not while Muybridge is there; he shot him once already." Afterward, when Muybridge had been removed "to another room" (probably the kitchen), they carried Larkyns into the lounge. Presumably he was dead by now, though McCrory told the court "Larkyns said nothing when we picked him up."

There was no cross-examination—no Perry Mason-like attempts to pick up on the apparent sudden knowledge of who the stranger was when McCrory claims he said, "Not while Muybridge is there." But then, no one was trying to show that Muybridge wasn't the killer. Again we have Larkyns charging through the house, but with Muybridge stopped part way through as he pursued him. Either the medical men were wrong, or these two witnesses had concocted a story that was at odds with reality.

The last of the direct witnesses to the killing was J. M. McArthur. Here once again there was a moment of Lewis Carroll-like distraction in the legal proceedings at the start of his testimony, particularly in the condensed style of the *Napa Daily Register*. McArthur started by saying that he lived in Oakland, but he was stopping at the Yellow Jacket mine, and the mine was in Napa County.

> [Here the Judge enquired if there was any doubt as to the mine being in Napa county. Mr Pendegast thought not—the Surveyor so reported it.]

This reads more sensibly, if less entertainingly in the *San Francisco Chronicle*, which explains why this matter was brought up. There was some doubt as to whether the ranch house fell in Napa or Sonoma County. If it had been the other side of the boundary line, the defense could have picked up on this to query the validity of the trial. But, as the *Chronicle* reported, "the policy of the defense has been to avoid all technicalities and no murder case has ever been tried with fewer legal objections or technical quibbles. To this fact is due the excellent progress made in the trial."

It might seem that this is a criticism of the defense team, but in fact King and Pendegast seem deliberately to have been playing the honesty card. Once McArthur was properly allowed to give evidence, he gave a view of the incident from inside the house:

> I did not see the shooting; was sitting in the sitting-room about half past ten in the evening. Larkyn and myself were talking. Murray spoke and said a man wanted to see the Major. I had previously heard a buggy drive up. Prickett's, I think it was. Spoke up and said let him come in. The Major walked to the door. A voice said "Good evening, Major." The Major said "It is so dark I cannot see you." A name was given, which I did not catch, but I heard the words "Here is a message for you from my wife." Then a shot was fired and the Major came into the kitchen with his hands upon his breast. He said, "Let me out, please." He passed out and fell.

It's not entirely clear whether McArthur is saying that Larkyns collapsed in the kitchen, or passed out of the house, then fell. McArthur was now to go into full Wild West mode:

> Muybridge came in, apparently following Larkyn. He came to the sitting-room door with pistol still in hand and raised. I covered him with my pistol and commanded him to halt. He halted. I said "You have shot the man, now give me your pistol." After some hesitation he gave it up. About this time Stewart got up, the men looked after Larkyn and Muybridge was taken into the dining room. Soon after I was told the Major was dead. We got up a team and took the prisoner to Calistoga.

It sounds from this as if McArthur had his pistol strapped to his side as we've come to expect from Western movies, but in a different newspaper report we hear the additional evidence. In fact, McArthur ran to his bedroom as soon as he heard the shooting and took his revolver from under his pillow. Muybridge then entered the house, pistol in hand, and as he passed the bedroom door, McArthur darted out, covered Muybridge with his pistol and got him to surrender.

The pistol used in the killing—a Smith and Wesson six shooter—was shown to McArthur in the courtroom, and he identified it as the one he took from Muybridge. It seems that the gun had been left exactly as it was found with four bullets remaining in the chamber as the report laconically commented:

> Witness was handling the pistol when Mr. Pendegast said, nervously: "We would rather admit that this is the pistol than have it slung around here."

His colleague Mr. King thought it might be better to have the thing disabled and asked the sheriff to remove the charges from the pistol, but Pendegast simply wanted it out of the way and hurriedly agreed: "That is the pistol. There is no dispute about it. Put it up." The pistol was removed, and the court (and particularly the understandably nervous Mr. Pendegast) was able to get back to the serious business. McArthur gave evidence about the gun that fits with the story that Muybridge fired into the air on the way up to the ranch:

> Think there was but one shot fired. I thought at the time there was but one barrel emptied, but the Constable afterwards informed me there were two barrels empty. We took a team, and took Muybridge to Calistoga. There were four of us in a two-seated wagon, one man drove, another held a lantern. Muybridge was on the back seat and I guarding him.

This was obviously an opportunity to get a feeling for Muybridge's state of mind and attitude, and Spencer, the district attorney, wanted to know just what happened, asking, "Did he say anything to you on the way down to Calistoga regarding the shooting?"

Pendegast objected, as he felt that confessions made after Muybridge was taken into custody should not be given in evidence, but the judge went with the prosecution and McArthur was allowed to continue. He seemed to find the calm certainty of Muybridge's workmanlike approach to committing murder unsettling. "He said he made all his arrangements before leaving San Francisco, that he didn't know but he might be killed in the affray or lynched by the miners afterwards, or something of that kind, and he had arranged all his affairs so that if he never returned his business would be all settled. He said he intended to kill Larkyns. He expected to find a camp of miners here, and miners were a pretty rough lot, and he did not know what the consequences would be, but he had resolved to shoot the Major and take the consequences."

McArthur went on to relate a rather quaint query he claimed to have overheard. A Mr. Searles apparently stood guard over Muybridge, sharing the duty with McArthur. Searles asked Muybridge why he came in the night, "frightening everybody," rather than in the daytime. Muybridge, Searles claimed, confessed he had done this to make sure that Larkyns could not shoot *him*.

This allegation that paints Muybridge in a cowardly fashion, lurking in the shadows when he could have called out Larkyns in the cold light of day, seems at odds with the truth. He had rushed to find Larkyns as quickly as he could and it was this driving sense of urgency rather than any calculation of his survival chances that was behind the timing of the attack.

McArthur was subject to more cross-examination than the other witnesses. The defense team asked for more detail of what Muybridge said and did. It seems likely this was intended to bring out some good character information that hadn't come through in the answers the prosecution had looked for. McArthur began by admitting that Muybridge had been sorry for any distress he caused at the house:

He made excuses to the ladies and begged their pardon for frightening them as he had done. He told them that he considered that this man Larkyns had destroyed his happiness or something like that.

Pendegast tried to bring McArthur round to admitting that Larkyns had provoked the attack with a superbly leading question: "Did he say, as one of his excuses, that "this man has seduced my wife"?

Judge Stoney was not going to let him get away with that, although his objection was not the most obvious one: "I object to that. We have not asked anything about that."

The trial judge seems not to have intervened, as McArthur carried on as if the prosecution had said nothing. He admitted that while Muybridge didn't use the words "seduce my wife," he had implied it in what he said about Larkyns destroying his happiness.

Then there was the minor matter of Muybridge failing to take the opportunity of an accident on the way to Calistoga to make his escape. In the excitement of the events, according to McArthur, "the boys ran off the bridge and I was thrown out [of the buggy]." But Muybridge made no attempt to escape despite not being covered by McArthur's gun, and even "expressed sorrow for the accident."

The prosecution was not happy with this and questioned McArthur further. He admitted on inquiry that the prisoner was tied and could have made no successful attempt at escape if disposed to do so.

Rather bizarrely, toward the end of McArthur's cross-examination, the two prosecution lawyers appeared to be operating at cross purposes. The district attorney had been asking McArthur a number of questions about his conversation with Muybridge and had come round to asking McArthur what (if anything) Muybridge had said about his wife. The district attorney asked, "Did he say where she was living at the time of the shooting?"

Not for the first time in this exchange, Pendegast objected: "It is immaterial if he did."

Judge Stoney—the second prosecuting attorney—now came back with the remarkable statement, "I think the whole conversation is immaterial, and the Court will instruct the jury so."

After which "the Court" is reported as saying to the witness, "That will do, Mr. McArthur." It seems rather presumptuous of Stoney, who was only acting as assistant prosecutor, to instruct the court in this high-handed manner. It is just possible he forgot where he was and decided to stop the line of argument his colleague the district attorney was pursuing because he thought it was doing them no good.

Equally possible, however, is that the *San Francisco Chronicle*, despite being generally one of the more accurate of the newspapers of the time, might have become slightly confused by the unusual number of judges involved, and it could instead have been the trial judge, Judge Wallace, who took charge at this point. Sometimes the newspapers used "the Court" to refer to the judge himself, but here, if it was indeed Wallace who complained, it's likely that it was the clerk who told McArthur to stop.

The final witness for the prosecution was the young driver, George Wolfe, who it had originally been thought would not be able to testify, as he had contracted the then life-threatening disease of measles. The *Chronicle* observed that "he came into the Court-room and the spectators cleared the passage instantly."

It seems likely up to this point that witnesses had been forced to push their way through the onlookers crowding around the door, but as it was widely known that Wolfe had just suffered from measles, he was given a wide berth. The young man described his journey to the mine with Muybridge:

> [Muybridge] asked me if anybody was likely to stop us on the road. I said "No; have drove three or four years and never been stopped." He said he had been stopped several times while travelling, and asked if it would scare the horses if he fired his pistol. I said no, and he fired. We went to the house at the Yellow Jacket mine. I saw him shoot Larkyns there.

Exactly what Muybridge had in mind when he asked about being stopped isn't clear. The *Napa Daily Register* version of his testimony assumes he meant stopped by robbers, though we have no record of him ever being held up. Wolfe then went on to describe, as the others had, how Larkyns had run into the house "and Muybridge after him."

There is one possible explanation for the difference between the witness accounts of Larkyns running through the house before dying and the medical evidence that suggested he would have collapsed on the spot. The prosecution wanted to show Muybridge as a cowardly killer who shot Larkyns before he could respond; however, from Wolfe's account it could be that Larkyns ran for it as soon as he heard Muybridge. Muybridge could then have pursued him through the house and shot him by the oak.

After Wolfe's testimony, proceedings were handed over to the defense.

When Cameron King opened for the defense, he made no attempt to deny what had happened, but rather attempted to show both that it was a justifiable act and that Muybridge was not in his right mind at the time. He began by attacking Larkyns's character, something that perhaps was necessary after the praise heaped on Larkyns when he died. King said that he would prove that Larkyns was a man of bad character—that he pursued Mrs. Muybridge with the deliberate purpose of seducing her. What's more, he suggested, Larkyns had rented rooms at Montgomery Street, representing himself as a married man, and had induced Flora Muybridge to accompany him there, a fact that he had boasted of to his friends.

This might seem a fairly innocent attempt to explain the background to Muybridge's mental state at the time of the attack, but the prosecution, probably accurately, reckoned that King was out to turn the jury against Larkyns, to make his death seem more acceptable. They weren't silent long—but King took some stopping. As the *San Francisco Chronicle* put it:

> When he referred to the character and previous history of Larkyns, counsel for the prosecution objected, on the ground that he should not be allowed to state matters which would not be admissible as evidence. The objection was sustained, and after several attempts to evade the ruling a peremptory ruling was made forbidding his stating such matters.

King now changed tack, going all the way back to the terrible head injuries Muybridge had suffered in the coach accident. The attorney suggested that Muybridge had never been normal since, and that his condition had been aggravated by Larkyns's terrible misuse of his trust and his wife. Although the science of the time wasn't capable of explaining it, this suggestion from Muybridge's attorney seems quite plausible. The psychologist Arthur Shimamura published a paper in

2002 that suggested Muybridge suffered damage to an area of the frontal lobe of the brain known as the orbitofrontal cortex. Shimamura suggests that both Muybridge's extremes of behavior and obsessive work on motion could be explained by his injury, as such changes in personality are not uncommon with this kind of damage. However, it should be borne in mind that any statements made about Muybridge's behavior at the trial were made with the intention of influencing the jury and may not have entirely reflected his true condition.

King's opening speech was impassioned, and in the great tradition that would later be upheld by fictional defenders all the way up to Perry Mason and Rumpole of the Bailey, he was not too worried about sticking to the words that had been heard. As the *Napa Daily Register* put it:

> He quoted poetry and read from authors on insanity. Mr. King took great latitude in his "statement," and was frequently checked by the Court and complained of by the counsel for the prosecution.

A precognitive ghost of Mason and Rumpole indeed. In the rest of his introduction, King stressed the unusual situation Muybridge was in. To punish Muybridge, he suggested, would make him a martyr, as he had cleared the world of a villain. He didn't make any attempt to deny the fact of the killing, but justified it as a just and right act, undertaken when Muybridge's frame of mind made him irresponsible. He showed the jury a picture of Flora (accompanied by complaints from prosecution) and the picture of Floredo inscribed with "young Harry" in Flora's handwriting.

It isn't entirely clear from this whether King was showing the jury Flora Muybridge's photograph to demonstrate what it was that Muybridge had lost, or to show the sort of woman she was. His closing words were superb, passionate oratory worthy of Shakespeare, appealing not to logic but going straight for the guts of the all-male jury:

> I ask you, if these facts be so, was this man guilty of crime? I do not believe he could be. Who is the man—even though he be of the soundest mind—that can say he would have acted differently? The love which every right-minded, honorable man bears his wife is the highest, best and noblest virtue. He is not a man who would not sacrifice himself to save his wife from dishonor and shame. And, I assert, that he who would not shoot the seducer of his wife, aye, even if he were to suffer ten thousand deaths, is a coward and a cur. Better, far better, death than that the seducer should

boast his conquest and his wife's dishonor with drunken companions over the flowing bowl, and point out the wretched man who walks the streets—a cuckold.

King finished conveniently for the lunchtime adjournment. Before any witnesses could be called in the afternoon, there was a minor hiccup thanks to a Mr. Klam, one of the jurors, who failed to turn up, but a deputy sheriff was sent out and soon managed to round him up.

Perhaps surprisingly, the first defense witnesses were Susan Smith and her daughter Sarah—the allies of Flora in her affair with Larkyns. The *Napa Daily Register*, which had failed to make any comment about the appearance of any previous witnesses, seemed to take a salacious interest in Smith:

> Mrs. S. is a woman of more than medium height, plainly dressed with a flower-trimmed straw hat. She gave her testimony readily and unconfusedly, and seemed to have about as clear a view of the situation as the Attorneys themselves: several attempts to confuse her failing miserably. She talked with vigor and at times rose with excitement of her story and delivered it with considerable dramatic effect.

The midwife stressed again and again how dramatic Muybridge's reaction had been to the revelation. She described him gasping, staggering, crying, collapsing—and simply, calmly she declared that to the best of her opinion he was insane at the time she last saw him before Larkyns's murder. By contrast with her evidence Muybridge sat impassive, as he would through most of the trial, seemingly past caring about anything that could be said.

Susan Smith seemed to enjoy enticing the court with the lurid details of what she had seen. As well as describing word-for-word the eventful meeting with Muybridge when he had realized the truth, she told more details of life with the adulterous couple. She mentioned that Larkyns "had told Mrs. M. he was of noble family in England, and intended to take her back and present her as his wife, and the baby as his child."

According to Smith, Flora's response was to say that it was too bad to treat old Muybridge so, but she loved Harry. Once, she told the court, Larkyns called in the afternoon while Mrs. Muybridge was still in bed. "I went into the room, and she lay on the bed with the clothes down to her waist, and Larkyns sat on the bedside."

Scandalous stuff indeed, and the *San Francisco Daily Morning Call* primly reported that "much more of this kind of testimony is omitted as unfit for publication."

The *Napa Daily Register* is less reserved and tells us:

> Mrs. S. had seen Larkyn sitting on Mrs. M.'s bedside when Mrs. M. lay exposed in a manner which the witness pronounced indecent, her breasts exposed, and Larkyn's hand resting on her shoulder.

And still Muybridge showed no flicker of emotion.

Sarah, the midwife's daughter, was, of course, not present to witness Larkyns and Flora together, but could tell the court of his qualities as a husband, and how his eccentricities were always showing through, emphasizing his unusual nerviness and the way he could become worked up about little things. According to Sarah, "Muybridge was kind and affectionate to his wife; gave her plenty of money whenever asked." She went on to say that he was "always kind and indulgent. Muybridge is very eccentric and peculiar, easily excited and very nervous."

Next on the stand was his friend and colleague William Rulofson. He put a lot of effort into establishing Muybridge's unstable character—so much so, in fact, that Muybridge took it personally. It didn't matter that this evidence was being given in an attempt to save Muybridge from the gallows; he was mortally offended by the way Rulofson attacked his behavior and suggested that many of Muybridge's actions were frankly not those of a sane man.

For example, Rulofson pointed out, Muybridge had actually turned down work on a number of occasions for the simple reason that he didn't particularly want to do it. With his strong Scandinavian work ethic, Rulofson could not accept such a wanton waste of earnings—where Muybridge simply saw it as building and preserving his image as an artist rather than a jobbing photographer. What's more, Muybridge was lax about pursuing debt.

> People owed him who did not pay, and he would never take the trouble to collect it. Judge Crocker, for instance, of Sacramento, owed him, I think justly, $700, which he made objections to paying. Muybridge at once proposed to send him a receipted bill; and so with others.

Perhaps most damning, Rulofson suggested, was Muybridge's behavior while photographing in Yosemite. As a picture by Muybridge's assistant proved, during one of his trips, Muybridge had sat on a rock, his legs dangling casually over the edge as if he were perched on a small boulder in the middle of a field. His position suggests that at any moment he might leap forward off his perch. Yet the picture also makes it plain that the rock in question juts out over a huge chasm; the drop below Muybridge's idly swinging legs plunged over 3,000 feet. Surely, Rulofson argued, this was not the action of a sane man.

Rulofson really rubbed it in—and perhaps this is why his performance so rankled Muybridge (though even now he showed no sign of his reaction in the courtroom). When the prosecutor, Judge Stoney, asked if he had given all the reasons he had for a belief of the defendant's insanity, Rolufson replied that to tell all would occupy, he judged, about two years. He went on to say that he was perfectly willing to devote this time to the cause if the counsel and jury were inclined to hear him.

This proposition, the *Napa Daily Register* dryly observed, was declined with thanks. Other business colleagues piled in to emphasize Muybridge's eccentricity and instability. Editor John Wentworth, fellow Mercantile Library board member Joseph Eastland, and even Silas Selleck, Muybridge's original inspiration for starting in the photographic business, told of his strange behavior. Most of them emphasized that before the stagecoach accident in 1860 he had been "genial," not to mention "pleasant and agreeable," but that afterward he was "irritable," "surprising and wavering," "more careless in his dress," and "not so good a businessman."

The defense evidence straggled to an end with the recalling of Smith, who claimed a right to be heard to "repel imputations upon her character made by a question from the District Attorney." It seems that Spencer had asked whether she was now living in open adultery, and Smith felt she needed to clear her name. The newspapers avoided any detail (the *Napa Daily Register* coyly remarks, "It soon became evident that something might be expected from a feminine point of view"), but it seemed to be an attempt to reduce the value of Mrs. Smith's testimony.

Despite his counsels' apparent closure the day before, Muybridge himself took the stand the following day, February 5th, "Day 3 of the

Muybridge Trial" as it is headlined in contemporary newspapers. Once again there was a brief delay for a naturalization hearing ("Court again manufactured a citizen out of foreign, raw material," as the *Register* put it).

Muybridge's lawyers had no doubt discovered just what Muybridge thought of Rulofson's testimony, and were careful to limit his comments to historical fact. Muybridge told of his accident, how his previously perfect health had since been troubled with headaches, and nothing more. Rather surprisingly, the prosecution elected not to cross-examine. But they were not finished with Muybridge's story. Instead they were to make every effort to show that this attempt to demonstrate temporary insanity was a futile one.

∞∞

Several witnesses who had spoken to Muybridge just before and after the killing were called to show how calm and unemotional he was. This was no raging madman, the prosecution suggested, but a cold and calculating murderer.

The first witness who was supposed to be called at this stage was J. M. McArthur, the man who guarded Muybridge on the way to Calistoga and chatted with him at length, but when he was called, as the newspaper report succinctly puts it, "No answer. Searched for— not found."

Next to be called back was George Wolfe, and for a moment the prosecution seemed about to collapse into farce as once again there was no answer. But Wolfe, the young driver who saw Muybridge before the killing, was finally located. He thought there was nothing unusual about Muybridge's behavior. Judge Stoney then went a little far in his enthusiasm and asked Wolfe about "the deportment of the prisoner the next day at Calistoga." An objection from the defense team was sustained, and though the point was "argued at some length between Judge Stoney and the court," the prosecuting judge was not allowed his own way.

The officials who had charge of Muybridge immediately after his arrest all commented on how cool he was, which, oddly, was taken as evidence of his normality, even when one of them commented that he was "much cooler than I should be if I had just killed a man, or than I think most any other man would be."

But the prosecution didn't get it all their own way. Rulofson kept up his certainty of Muybridge's unsettlement, describing outbursts of grief, and the sheriff was equally able to describe Muybridge's unusual excitement when he heard that the district attorney had been to see his wife to try to get her to testify against him.

The trump card in the prosecution's case was the evidence given by Dr. G. A. Shurtleff of the Stockton Insane Asylum. Shurtleff had in fact already been called by the defense, to the prosecution's surprise and complaints, in an attempt to prove that Muybridge's coolness on the following day was a sign of madness, not normality. Shurtleff dodged the issue, saying that he could not rely on evidence of Muybridge's state given by laymen.

When now called by the prosecution as an expert witness, Dr. Shurtleff was able to state that he had seen or heard no evidence that would lead him to believe that Eadweard Muybridge was insane. He believed that Muybridge understood what he was doing and its conse-quences. This was, he said, no madman. It seems he could rely on the evidence of laymen to decide that Muybridge wasn't mad even if he couldn't rely on such evidence as proof of insanity.

Shurtleff's evidence went on for some time, with the defense work-ing through various books on mental illness, trying to find parallels with the Muybridge case that Shurtleff would not support. As the *Napa Daily Register* put it:

> Examination of the Doctor continued to considerable length, in the read-ing of various cases and authors and a request of the Doctor's opinion on each. The occasion gives the reporters a long rest, who improve the oppor-tunity to lean back, stretch their legs, and sharpen up a fresh supply of pencils.

By the afternoon it was time for summing up. Once again the prosecution was taking no risks and left this delicate task to Judge Stoney. He spoke simply and effectively. He tried to make it clear that unless the defense had proved that Muybridge was insane—something Dr. Shurtleff's testimony made highly unlikely—then Muybridge was guilty of first-degree murder. This might seem obvious now. He had committed the crime; it was premeditated. Yet it was not uncommon in Western courts at the time to accept that to kill your wife's lover was a reasonable and acceptable response. Stoney made a very impressive job of putting the decision as a simple one to the jury—either Muybridge was mad or he was guilty.

Stoney also tried to undo the damage that Cameron King had done in attacking Larkyns's character. He said that he was

> surprised that King should undertake to vilify and abuse Larkyns, because it was gratuitous. It was following him into his grave. There are friends of Larkyns here, and at their request I repudiate the imputation on his memory. There is no evidence before this Court that Harry Larkyns ever was improperly intimate with Mrs. Muybridge. All the testimony on that subject was given for another purpose, and should not be accepted as evidence of that state of affairs. The adulterer is not an outlaw.

It was early evening, after nearly two hours of recess, by the time Cameron King began to speak for the defense. If anything, the crowds were even worse. According to the papers, the courtroom was thronged long before the appointed hour for re-assembling, the adjacent jury room was filled with spectators, and so great was the pressure that in a series of skirmishes some of the reporters' chairs had vanished.

King tried to cast doubt on Dr. Shurtleff's evidence, pointing out that the majority of people who actually saw him at the time thought Muybridge unhinged. But, clearly aware that a plea based on his long-term insanity was shaky, King decided instead to concentrate on his client's state of mind that day, to suggest that his sanity was certainly disturbed by the shocking meeting with Susan Smith. What's more, whatever the law said, Muybridge had the moral right behind him. King quoted the Bible, suggesting that the passage he read was reflecting exactly such a passionate response, a response we would now consider temporary insanity: "The man that committeth adultery with another man's wife, even he that comitteth adultery shall surely be put to death."

He went on to praise Muybridge's character with a real tug at the jury's heartstrings and Victorian sentimentality. And to get a little dig in at the use of Judge Stoney on the prosecution side:

> Spoke of Stoney as "the hired counsel" and questioned the good taste of his appearing for the prosecution. The District Attorney selected "by the voice

of the people for his learning and genius" should have been sufficient for the task.

But the blockbuster effort was left to Wirt Pendegast. In a two-hour speech that left the listeners awestruck, he called for justice over and above any detail of the law. Pendegast began by reflecting back the prosecution's presentation of the decision as a simple choice, but with a different spin. Muybridge was, he said, guilty of first-degree murder or of nothing at all. He should hang or go free; there was no middle ground.

> I agree with Judge Stoney that the prisoner at the bar is guilty of murder in the first degree, or he is guilty of nothing. Either send him from the Court-room a free man, or send him to the scaffold. He deserves absolute free-dom or he deserves death. Between the two ye are the judges.

In effect, Pendegast had decided to raise the stakes in the poker game over Muybridge's future. They weren't going to play around for any reduced sentence for temporary insanity—it was not guilty or be hanged.

Yes, said Pendegast, Muybridge killed Larkyns—and this was no self-defense, whether you take that to mean keeping himself alive or defending his name and "property." Yet the fact was that Muybridge had loved his wife, was devoted to her, and had received the ultimate shock.

Now came his masterful attempt to show that justice was more important than any pile of law books. The prosecution had said that there was no law that allowed such retribution, but did there need to be? It was, he said, provided by a higher law. Pendegast compared Muybridge's response to a mother giving nourishment to her child. There was no legal requirement for this, but such cases were covered by the higher law of God and nature.

Technically, logically, Pendegast's argument may seem entirely wrong. Perhaps appealing to the "law of nature" is leaving things to survival of the fittest. Isn't our whole modern civilization—and the Christianity that Pendegast seemed to be appealing to—based on our ability to rise above nature? Yet such was the power of his oratory—particularly with a jury that was well versed in the basics of vengeance and nature—that logic didn't really enter into it. As Judge Stoney later observed, Pendegast managed his feat "not by perverting the facts, or by distorting the law, but by raising the minds of the 12 men whom he addressed, above the influence of the law and the facts." We might now say that he took their minds below rather than above reality, but the sentiment remains powerful.

In building the picture of an action justified in all but the letter of the law, Pendegast showed how he believed the law itself was lacking. There was such a thing as justifiable homicide, as when the killer would lose his own life otherwise. Not to mention the acceptability of killing in times of war. He took the jury through the cases when taking another's life was acceptable in the law. Some of these cases, he argued, were much less valid than the right to punish a seducer.

Pendegast ended with a violent assault on the jury's emotions.

> When [Larkyns] entered the front door as his trusted friend, and stole from him his rarest jewel, the love and good name of his wife, and writes "prostitute" across her brow, and dishonors and ruins the home of her husband, that husband is to ask the law to protect him from a repetition of such conduct. . . . I cannot ask you to send this man back to a happy home—he hasn't any—the destroyer has been there and has written all over it, from foundation stone to roof tree, "Desolation! Desolation!" His wife's name has been smirched, his child bastardized, and his earthly happiness so utterly destroyed that no hope exists for its reconstruction.

He certainly got through to Muybridge himself, who for the first time in the trial was showing his feelings. His entire body was shaking, his head held tight in his hands. At least, Pendegast finished, they could allow him to go back to his profession "upon which his genius had shed so much luster." His last words made it clear to the jury that a verdict of "not guilty" would not only give Muybridge life but also be appreciated by California as a whole.

> Do this, but this, and from every peaceful household, and from every quiet home within the state there will come to your verdict the echo of a solemn and deep Amen!

It was by now half past nine at night. Such was the strength of the speech that according to the *Napa Daily Register* report, "Many were surprised from their propriety into heartily applauding." The *San Francisco Chronicle* reports that this was "one of the most eloquent forensic efforts ever heard in the State," which resulted in a "storm of applause."

Moved though the judge may have been by the speech, the applause did not go down well. The *Chronicle* went on to say that Judge Wallace ordered the immediate arrest of those causing the uproar:

> His eye caught one of the applauders, Dan McCarthy, landlord of the De-
> pot House, and he ordered the Sheriff to take him into custody and any
> others he could find. He found no others. The Judge said he would attend
> to McCarthy after the case was over, and would clear the room if any more
> demonstrations were made.

After the excitement had died down, District Attorney Spencer gave a final summing up for the prosecution. He had to admit he was in a difficult position. It was an embarrassing position to follow the eloquent Mr. Pendegast, but he would discharge his duty to the best of his ability. He knew that his position was unpopular, but he had a duty to perform and trusted in the law. This wasn't the position of an over-confident man. He began by attacking the possibility of madness, based on Dr. Shurtleff's denial. The only witnesses, he pointed out, to support insanity were Rulofson and the midwife, Smith. The former he dismissed as a friend, the latter as an unreliable witness.

Next, Spencer tried to show that Muybridge's action had been anything but righteous, denouncing the shooting in the dark as murderous and cowardly. He made it plain that he believed Larkyns was more sinned against than sinning. In the end, Spencer claimed, the defendant knew what he was doing—he knew right from wrong.

Finally, it was time for Spencer to take on the most dangerous aspect of the defense, the attempt by Pendegast to suggest that justice was more important than the law, and that natural justice had been done in this case. He pointed out that Pendegast had been in the legislature for four terms, and presumably could have pressed to change the law if he really thought it inadequate. After making a comparison with previous cases, Spencer attacked the biblical basis for Pendegast's argument:

> Produced at last a bible, which counsel endorsed. Bible did not say an adul-
> terer was to be put to death without law. Read also the case of the woman
> taken in adultery, to prove the veniality of that offense. There is nothing in
> the bible to suggest that the husband shall take the law into his own hand;
> Christ only said "go and sin no more." The virtue of women rests not on

their husband's revolvers but in their own purity. You have no right to ignore the law, and must find the prisoner guilty unless you conclude he is insane.

Judge Wallace gave the jury their instructions. And within 45 minutes of Pendegast's last, emotional plea they were in the jury room.

The minutes ticked away. Once a foreman had been elected, the jury immediately took a ballot. At this stage seven were ready to acquit Muybridge while five wanted him hanged. One juryman—it wasn't identified which side he was aligned with—took off his coat and folded it to make a pillow. According to the *Napa Daily Register*, he then told his fellow jurymen to wake him up when they agreed with him, swearing to "stay there until Resurrection day rather than to acquiesce to any other view." The rest of them, the paper informs us, did not debate the trial much more that night but instead "discussed currency, taxation, fees and other relevant questions." With only three-quarters of an hour to midnight, it was decided to adjourn to the next day.

Another ballot the next morning came up with the same result. In discussion it came out that pretty well all were actually in agreement on whether Muybridge was justified in his actions; it was the matter of insanity that underlay the split. According to one report, one of the jurors now asked every married man present to think of his own wife. What was she doing right now? Could he be sure that she was faithful? Another man might be in bed with his wife this very minute.

Whether or not this appeal to the paranoid really took place, to everyone's surprise, when they took a third ballot, the result was unanimous. In all it was around 13 hours before they returned to court. Muybridge was brought up from his cell, now seeming once more impassionate.

The foreman was asked if the jury had reached a verdict that they all agreed on. He replied that they had. Instead of announcing it, he passed a slip of paper to the clerk, who checked it and passed it on to the judge. The silence was total. The sounds of the paper being opened, of the judge's clothes as he moved were audible throughout the courtroom. The judge nodded. "Record the verdict," he announced.

With what now seems an unnecessary drama, the clerk made a note of the verdict. The crowd, the attorneys, and Muybridge himself were still unaware of the outcome of a trial where his life hung in the balance. Finally, clearing his throat, the clerk rose:

> Gentlemen of the jury, listen to your verdict as it stands recorded. "The People of the State of California versus Eadweard J. Muybridge. We the jury find the defendant . . . not guilty.

Muybridge collapsed, only prevented from slumping to the floor by Wirt Pendegast, who caught him as the freed man sobbed uncontrollably. The *San Francisco Chronicle* pulled out all the stops describing the scene:

> At the sound of the last momentous words a convulsive gasp escaped the prisoner's lips, and he sank forward in his chair. The mental and nervous tension that had sustained him for days of uncertain fate was removed for an instant, and he became as helpless as a new-born babe. Mr. Pendegast caught him in his arms and thus prevented his falling to the floor, but his body was limp as a wet cloth. His emotion became convulsive and frightful. His eyes were glassy, his jaw set and his face livid. The veins of his hands and forehead swelled out like whipcord. He moaned and wept convulsively but uttered no word of pain or rejoicing. Such a display of overpowering emotion has seldom, if ever, been witnessed in a Court of justice.

In the typical dramatic style of the time, the words "a display of overpowering emotion" appear in the newspaper in capitals as a headline. This wasn't the end of it. Pendegast got his client into a chair, but Muybridge's reaction of absolute shock continued:

> He rocked to and fro in his chair. His face was absolutely horrifying in its contortions as convulsion succeeded convulsion. The Judge discharged the jury and hastily left the Court-room, unable to bear the sight and it became necessary to recall him subsequently to finish the proceedings. The Clerk hid his face in his handkerchief. Mr. Johnston and the Prosecuting-Attorney were compelled to leave the room, and some of the jurors hurried away to avoid the spectacle. Others gathered around to calm the prisoner, and all of them were moved to tears. Pendegast begged Muybridge to control himself and thank the jurymen for their verdict. He arose to his feet, and tried to speak out but sank back in another convulsion. He was carried out of the room by Pendegast and laid on a lounge in the latter's office.

This Mr. Johnston, who had presumably helped Pendegast with the limp ex-defendant, is then reported as saying sternly, "Muybridge, I sympathize with you, but this exhibit of emotion is extremely painful to me, and for my sake alone, I wish you to desist."

This appeal seems to have got through. Muybridge suddenly sat up and said, "I will, sir, I will be calm, I am calm now," and then, according to the reporter, "his emotion subsided, so that in a quarter of an hour he was able to go upon the street."

He was to find that Napa had turned out in force. He was feted and given a free lunch; he returned to San Francisco that afternoon a free man. According to the *Napa Daily Register*, he was to leave that night for San Francisco, where, since he was so confident of acquittal, "he had an engagement to dine with a friend tomorrow."

All the evidence points to a Muybridge who was anything but confident, and if this dinner engagement existed, it was more a result of wishful thinking than confidence. Perhaps the best last words on the saga are the simple summary in the records of the sheriff's office at Napa. The same worn record book has a brief entry on what appears to be a famous Western name, Virgil Earp, aged 27, who was booked for burglary. (*The* Virgil Earp, older brother of the famous lawman Wyatt Earp, did spend some time in California, but in 1870 he appears to have been in Missouri, and was a peace officer rather than a burglar, so this is unlikely to have been him.) Muybridge's entire entry, summing up such a dramatic part of his life, reads as follows:

Muybridge E. S. [*sic*]—Murder—Arresting Officer Cromwell—arrested September 19th

Tried Palmer Court—arraigned October 19th

Held to Answer—Discharged February 16th 1875

English—Citizen of California—Age 47

Artist

PART TWO

THE DISSECTION OF TIME

SIX

THE GREAT PANORAMA

The Fire Alarm Office has an important mission to perform, and is intrusted, in a very large degree, with the safety of property and, to no inconsiderable extent, of the lives of citizens. There are a Superintendent, three operators and two line repairers. The operators are on duty eight hours each out of the twenty-four, so that at all hours of the day and night a public guardian is constantly watching, with one eye on the instrument and the other on a book. The office is located on Brenham place, overlooking Portsmouth Plaza, in the third story of the Exempt Building, with two conveniently large windows, commanding a comprehensive view of the entire lower and business portion of the city.

San Francisco Daily Chronicle, February 11, 1877

Muybridge was free, released without a stain on his character despite indisputably committing murder. Psychologist Steven Pinker suggests that this reaction by the jury, not uncommon in California courts at this time, is a natural result of the way men are programmed to defend their wife to maximize the chance of sustaining their genetic code. In his *How the Mind Works* Pinker writes:

> Throughout the English-speaking world, the common law recognizes three circumstances that reduce murder to manslaughter: self-defense, the defense of close relatives, and sexual contact with the man's wife. . . . In several American states, including Texas as recently as 1974, a man who discovered his wife in flagrante delicto and killed her lover was not guilty of a crime. Even today, in many places those homicides are not prosecuted or the killer is treated leniently. Jealous rage at the sight of a wife's adultery is cited as one of the ways a "reasonable man" can be expected to behave.

This need to defend a mate reflects the unusual nature of the human animal when set alongside the other primates. Our tendency to monogamy is, despite a considerable enthusiasm for straying among a sizable percentage of the population, inherent and sits uncomfortably with the way we live in large groups. The only other primates that are monogamous (it's pretty rare) are those where each pair lives in a clearly defined territory, so there is little danger of the monogamous relationship being usurped.

According to zoologist Clive Bromhall in his book *The Eternal Child*, when the early prehumans moved from the trees to the savannah in response to climate change, they were forced to live in large groups in order to survive attacks from the ravenous predators they faced. Such group living would not have suited the mature prehumans any more than it does today's chimpanzees, who will rapidly destroy each other in a blood bath if forced into groups of larger than a handful. The characteristics that natural selection emphasized in prehumans that made it possible to live in large groups also resulted in physical changes that involved keeping more of the physical characteristics of a baby. This, Bromhall suggests, explains why we have such delicate, easy-to-tear, unprotected skin, why we have an ungainly upright stance, large heads, and much more—they are all traits of the infant.

This mechanism, where breeding for ability to work cooperatively can have the side effect of producing an infantlike version of the animal, is something humanity has since applied repeatedly to its domestic animals. The dog has much more in common with a wolf cub than with the mature wolf it was bred from. In a fascinating long-term experiment between the 1950s and the 1990s, Russian geneticist Dimitri Belyaev selectively bred Russian silver foxes for docile behavior and showed just how early man managed to turn the wolf into a dog.

Over 40 years—an immensely long experiment, but no time at all in evolutionary terms—the fox descendants began to resemble domesticated dogs. Their faces changed shape. Their ears no longer stood upright, but drooped down. Their tails became more floppy. Their coats ceased to be uniform in appearance, developing color variations and patterns. The domesticated foxes spent more time in play, and constantly looked for leadership from an adult. As they became more cooperative, they took on the physical appearance and characteristics of overgrown cubs. Just as humans became overgrown babies.

Bromhall suggests that the same modifications that made us more infantile in appearance also tended to develop a more infantile, one-to-one relationship between male and female, more like that of a mother and child than a normal primate relationship with a mate. Just as our slow run and thin skin actually made us easier prey but were worth suffering to gain the ability to work and live together, so we gained monogamous behavior, even though it is hard to support in crowded environments. And from this unlikely combination of monogamy and the crowd of human society came the "natural" response that Muybridge would make to the discovery of his wife's affair.

However much the jury had justified his action, after the trial Muybridge had a strong urge to get out of San Francisco. He was to spend over two years away from the United States. It seems that while in jail he had planned on sailing to Central America should he be released, and less than two weeks after the trial's conclusion he was on a ship of the Pacific Mail Steamship Company (probably the *Montana*), en route to Panama, then part of Colombia, and on to Guatemala.

The outcome of his trip was a series of fascinating photographs of this then underdeveloped part of the world—but also a cleansing, as if the simplicity of life away from San Francisco could wash away the worst of his experiences. This period of his life might not have contributed anything to the furtherance of the Muybridge legend or to his scientific and technical achievements, but it seems to have been cathartic, to have given him the opportunity to redevelop his burning enthusiasm for his career and for life.

Muybridge's trip to Panama and Guatemala was partially sponsored by the steamship company, both to increase awareness of these destinations and to publicize the Guatemalan coffee industry, which had begun in a tiny way in the 1750s but was now beginning to challenge earlier coffee plantations, from the original Dutch sites in Ceylon (Sri Lanka) and Java to the later, more successful plantings in Brazil and Jamaica. Just a few years before Muybridge's visit, in 1860, total coffee exports from Guatemala were practically nonexistent. By the time Muybridge made his trip, more than 15 million pounds of coffee beans were being exported every year.

Planting coffee in Guatemala meant hacking into virgin rainforest. In the twenty-first century a newspaper report on the subject would more likely express horror at the destruction of the environment caused when it was cleared for the plantation than admiration of the achievements of the coffee growers, but in 1875 the *Panama Star* would report with enthusiasm that Muybridge had taken 20 views "illustrative of the cultivation of coffee in all its stages from the time the forest is cleared away to receive the seed, all along until the crop is ready to be embarked." According to a flyer put out by Muybridge in Guatemala, his visit was "due to the Pacific Steamship Company which has assumed the costs of this venture for the purpose of making the riches and beauty of this interesting land known abroad." This flyer was in Spanish and gave Muybridge's name as Eduardo Santiago Muybridge. Though the steamship company had apparently "assumed the costs," a lack of clarity over the exact relationship between Muybridge and his employer would prove a problem on his return.

Muybridge may also have thought it useful to get away in case Flora tried once again to get her hands on some of his assets as a part of a divorce settlement, but events would render any worry unnecessary. In his absence, on July 18th, only five months after the trial, Flora Muybridge died, with a convenience that might have seemed suspicious had Eadweard not been so distant. The newspapers reported her death with a mix of sensationalism and pity. "Poor Flora is cold," read one of the headlines. She had, it seems, been suffering a couple of weeks from "paralysis," though there was no indication in the newspapers of the disease that lay behind a diagnosis that is little more than a symptom. Her last words were reported to be, "I am sorry." She was 24.

According to the official records, the actual cause of her death was not paralysis but rheumatic fever, a delayed response to an untreated streptococcal infection of the respiratory system, most common in children, where the infection spreads to and damages the heart. She was originally buried in the "cosmopolitan" section of the Odd Fellows Cemetery, but was later disinterred and moved to Green Lawn Memorial Park in Colma, California, where she was reburied in grave S, tier

22, of the Odd Fellows section. (The Odd Fellows is an organization that originated in the United States in 1819, modeled on the Freemasons. This implies that Flora's father, uncle, or guardian was a member of the order.)

The *San Francisco Examiner* of July 19, 1875, took the opportunity to look back over her life in an obituary that seems excessively full for a part-time photo retoucher, if it were not for the dramatic story of her affair. Headlined "The Last Call—Death relieves Mrs. Flora Muybridge from a life of sin and shame," it carries a smattering of errors from the location of the shooting (given as Pine Flat instead of the Yellow Jacket) to attributing Muybridge's acquittal to "the technical grounds of insanity." The article also brings her young son, Floredo, back into the picture:

> Mrs. Muybridge was a beautiful woman, of more than average intelligence, and her sad fate was in keeping with her checkered career. . . . Her husband, from whom she separated after the killing of Larkyns, is in South America, photographing the scenery of that country. Her babe is with a French family at Mission, who have kindly cared for it during the illness of its mother, and will, it is believed, adopt it. Her friends and acquaintances in this city are making efforts to procure the means of giving her remains a burial befitting the station she formerly occupied in life.

Muybridge would never reinstate Floredo as his son, but neither did he entirely ignore him. Despite the newspaper article's suggestion, the boy was not still with the French family, but had been put into an orphanage. Muybridge found the family and paid their expenses. Floredo had moved to a Protestant children's home from the Catholic one the French family had placed him in. Muybridge did not take the boy in—he was still convinced that he was not Floredo's father—but he did allow Floredo to keep his surname, and visited him several times during the first few years. After that, though, he seems to have felt his obligation to the boy complete and Floredo was left to his own devices.

Muybridge would visit with his son for a last time around the time of Floredo's eighteenth birthday, when the then highly successful photographer was preparing for a tour of the Far East, but seems to have found little common ground with the young man. Whoever he took after, Muybridge or Larkyns, Floredo showed little of his father's energy. Early on, Muybridge had put some money aside for Floredo's education, but after their last meeting this seemed unlikely to be needed. With what now seems a callous gesture he changed his will, cutting Floredo out, bequeathing the money instead to two of his cousins back in Kingston-upon-Thames.

This was unfair to Floredo (or Flodie as he came to be known, a nickname that was not much better than his given name), who whatever the truth of the matter had been brought up to believe that Muybridge was his father, a belief that Muybridge never attempted to alter. In 1884, before his tenth birthday, Floredo was removed from the orphanage to be fostered at the Haggin ranch outside Sacramento. His foster parents were first the Bucks, David Buck being a leather worker who made harnesses for the ranch. Then, when Mrs. Buck divorced her husband and married Carsten Tietjen, also working at the ranch, Floredo moved on with her.

His life on the ranch was not unpleasant, but hardly conducive to any desire for a higher education. According to Frank Tietjen, Carsten's son, who was like a brother to Floredo after the remarriage, "He got his schooling on the ranch, and could read and write just fine."

Floredo never amounted to much and drifted between jobs, working as a gardener and handyman, and eventually simply fulfilling the role of town drunk in Sacramento. Milton J. Ferguson, state librarian at the California State Library, wrote of him in a letter to Janet Leigh (the daughter of Wirt Pendegast, Muybridge's defender in the murder case) on September 23, 1927: "We hunted him up in 1916. He came to the library several times and was a pitiful specimen of humanity. He disappeared soon after and we have been unable to locate him since."

They may have been unable to locate him, but Floredo was still alive when the letter was written. In his seventieth year, on February 1, 1944, he was killed by a passing car. An article in the *Sacramento Union* rather inaccurately filled in the details: "Floredo was killed at the age of 60 when struck by an auto at the corner of Fifth and I Streets, a year ago February 2. At the time of his death he was living with H. O. Adams, a tile maker, at 911 26th Street." The article was written over a year later, in May 1945, which may account for getting both the date of Floredo's death and his age wrong. It was a sad end to a largely nondescript life. The whole Muybridge saga had passed Floredo by unnoticed.

☙☙☙

The Central American trip proved a mixed blessing for Muybridge's fortunes, though it was a remarkable adventure, providing a distraction that must have distanced him from his trauma. The journey itself certainly had distractions enough. After stopping for six weeks in the whitewashed adobe surroundings of Panama City, taking expeditions into the wilderness and ruins around it, Muybridge caught a steamer to San Jose de Guatemala, stopping off en route to photograph a series of ports. From there he took an overland journey to Guatemala City, which was just as hazardous as, and much more exotic than, his trips from New York to San Francisco.

An English traveler Whetham Boddam-Whetham took the same route around that time and reported on the almost nonexistent roadway where "at one moment one's wagon was up to the axle-tree in deep ruts, at the next jolting over great rocks and stumps of trees." For Whetham, the people he saw along the way rather spoiled the view: "Here and there an open thatched hut peeped out from a group of banana and orange trees; picturesque enough in itself, but marred by the untidiness and unwashed appearance of its occupants."

Apart from the coffee plantations he was engaged to publicize, Muybridge found ample photographic subjects in vistas across Guatemala City from the nearby Cerro del Carmen mountain, and the ancient remains of Antigua, loured over by the stable but still active volcanoes that towered behind it. When he wasn't photographing the plantations or the ruins or the natural beauty, Muybridge was presented with arrays of soldiers, lined up in front of local monuments to be captured on the plate by the great photographer.

On his return he advertised portfolios of 120 photographs for $100, but there was little enthusiasm for this offer. Since Muybridge had so taken against Rulofson after his enthusiastic description of his friend's madness, these were not sold through his company (in fact he never worked with Bradley and Rulofson again), but directly through Henry Bradley, who had left the studio in Rulofson's hands. Muybridge was able for the moment to work through Bradley's address, keeping the benefits of his association without having to deal with the man who he felt had betrayed his good name in his trial.

That the Central American photographs weren't a huge success is hinted at by an advertisement Muybridge put out soon after, describing his experience, his flexibility (working "with plates from 24 by 20 inches in size to the smallest card") and emphasizing his willingness to photograph anything from railroads and private residences to copying pictures and drawings. Still no portraits, though. Interestingly there is no reference to steamship companies—perhaps not surprising, considering the difficulties that surfaced between Muybridge and the sponsors of his tour.

Muybridge had thought that the Pacific Mail Steamship Company had commissioned him to take the South American photographs, and was expecting a good-size payment in return. Unfortunately, the company had a very different picture of their relationship. They had given Muybridge free passage and felt that this was the limit of their sponsorship. And they had no interest in buying any pictures. Muybridge, not for the first nor the last time in his life, threatened litigation, but there was no evidence to prove that the Pacific Mail Steamship Company ever intended to pay him, and so he never took them to court.

He had, however, produced four magnificently bound albums from the Central American trip. One was sent to the steamship company as a sweetener alongside that never-to-be-paid bill. A second went to the widow of Wirt Pendegast, who had recently died leaving a young family, with the third to another lawyer (though oddly not to Pendegast's associate, Cameron King), and the final album was delivered to Mrs. Leland Stanford. Perhaps this was because Stanford had encouraged his friend Pendegast to work for Muybridge, but realistically it was also because Stanford offered a big possibility of future lucrative employment in a field that Muybridge found fascinating.

The album sent to Mrs. Stanford soon produced more fruitful payback than futile attempts to get money out of the steamship company. She and her husband had recently moved from Sacramento to an elaborate mansion on exclusive Nob Hill in San Francisco. (This is not the origin of the term "nob" to mean someone of wealth or social distinction, which dates back to seventeenth-century England. The site, more properly California Street Hill, was nicknamed because of wealthy occupants it was drawing.) Mrs. Stanford commissioned Muybridge to document her new house, and soon he had produced a series of 41 photographs that made the most of the Stanfords' dramatic, if ostentatiously nouveau riche, artistic tastes.

Traveling up Nob Hill to capture the Stanford house on glass was part of the inspiration for the most remarkable landscape photography Muybridge ever undertook. It is probable that being recently paid by Mrs. Stanford was another contributory factor, as his next project was not to be a cheap venture. Muybridge saw in the mansion of the Stanfords' next door neighbors, the Mark Hopkins house, a superb opportunity to undertake one of his favorite photographic specialties—the panorama.

Although panoramic views had been undertaken by painters, there had seemed something natural about the combination of a camera and a tripod that allowed the operator to gradually swing his lens around and capture a series of photographs which were then joined together to produce a 360-degree sweep of vision. Muybridge had tried this on a number of occasions, but now he was to capture uniquely the essence of 1870s San Francisco.

One of the features of the elaborate structure of the Hopkins house was a tower, and it was from the top of this that Muybridge took his view over the city. In the foreground were the elaborate houses of the other millionaires. Looking out to the northeast was the bay, several hundred feet below, with the tall ships clearly visible. The sheer vibrancy and newness of the city comes through in the images Muybridge captured, first on 8- by 10-inch plates in 1877, and then the next year on huge 18- by 22-inch negatives, making up a remarkable display that was more than 17 feet long.

These panoramic views were sold under the aegis of another gallery, that of George Morse. In the prospectus for the 1877 view, Muybridge gets quite carried away by his own brilliance:

> The photograph of San Francisco, recently made by putting together a succession of views which, taken from a commanding central point, make a complete circuit of the horizon suggests to us that among the many won-

derful features of our city the character of its topography is not the least
deserving of attention, although it has been overlooked, at least by people
generally, until Mr. Muybridge discovered and utilized the artistic value.

We shouldn't be too hard on Muybridge—there are few artists or
writers who haven't found it necessary to be effusive about their own
work when producing a "puff"—and the fact is that he wasn't really
exaggerating. The panorama is stunning and does give one a feel for
the lay of the land in the city that could never be appreciated from
individual photographs. The remarkable view brought the *Alta California* of July 22, 1877, to comment that "other States and cities may be
greater and in many respects more attractive, but none can equal California and San Francisco in the panoramic scenic effect."

The article goes on to give us a picture of what is involved:

> Let us imagine a small ant wishing to get a comprehensive view of a painted
> Japanese dinner plate. He would succeed if he could get a thimble upright
> in the middle of the plate, then climb to the top of the thimble and look by
> turn in every direction. The ant, in that hypothesis, occupies a similar po-
> sition to that of the man in San Francisco which represents the saucer, and
> the palatial dwelling of Mark Hopkins, on California Street Hill, is the
> thimble.

For $10 you could buy all 7 feet of the first, smaller panorama,
mounted on a cloth cover—but along with the larger version, it was
soon off the market. The early days of San Francisco were plagued by
fires, and in 1878 George Morse's gallery, where much of Muybridge's
material was stored, burned down. All the plates from his Central
American trip were destroyed (though luckily he had already issued
his four presentation albums) as were the negatives of both small and
large panoramas. He reshot the sequence, using the 18 by 22 plates,
that April.

As usual, Muybridge was very PR conscious. On the new panorama
prospectus is a coy instruction, asking the reader "after making a note
of its contents, please hand this circular to someone whom it will probably interest."

Another example of his marketing skill was the way Muybridge
had come to realize that one of the best ways to get your name known
in San Francisco society was to be popular with the wealthy wives. He

presented copies of the panorama both to Mrs. Hopkins, whose house he had used, and to Mrs. Stanford next door.

It is quite possible that his public relations effort paid off. Even during the taking of the panoramas he had done a little further work photographing the horse Occident for Leland Stanford, but soon after the success of the panoramas he was to take on a new contract. With a degree of commercial caution, however, he looked for a way to reduce the risk of having too much of his income tied in with the Stanfords. Toward the end of 1877 he made the remarkably modern-sounding suggestion to the city of San Jose that he should photograph its public records, replacing the laborious hand copying that was the norm at the time by making quick, compact, and accurate photographic records.

It has been suggested that the idea came to him when he put together a sales gimmick for his panorama—a single print of the whole panorama (in four sections) that also included a key identifying 221 locations, forming a miniaturized version of the information contained in the full-scale panorama. If Muybridge could compress his visual images this way, why not also with San Jose's public records? Muybridge proposed photographing at one-sixth of the original size, squeezing the records into a more manageable form as well as making the copying process much less laborious and liable to error.

In one sense Muybridge's idea was nothing new, and his reduction in scale by one-sixth was, frankly, unadventurous. Though there is no evidence Muybridge knew about it, an English scientist named John Benjamin Dancer had begun experimenting with miniature texts almost as soon as photography became possible. Back in 1839, he produced his first text reduced in scale a remarkable 160 times. There was already a passion for miniature books, handwritten with incredible precision—Dancer's interest was similarly in the entertainment value of the very small. In 1853, he began selling tiny texts as slides to be read under a microscope.

Initially, miniature photographs were considered nothing more than a novelty, though as early as 1853, suggestions had been made that microphotographs would be a good way to preserve library materials. Briefly they were in vogue as an early form of airmail—pigeon post—which was tried out in both the United States and France in

1871, sending thousands of letters on a tiny photographic print carried by a homing pigeon. Apart from these oddities, though, the primary business application of recording documents would not be practically achieved until the New York City banker George McCarthy built his Checkograph machine in the 1920s for making photographic copies of bank records.

Muybridge's predecessor to McCarthy's successful device was never put to the test—the San Jose Board of Supervisors turned down his offer for two reasons. First, they argued, the text would be too small. Some of the documents had scribbled remarks in the corners of the pages or between lines that would not be readable at that scale. And, decisively, Muybridge's estimate of the cost, $5,892.25, which worked out to 35¢ a page for the 16,835 pages he would have to copy, was too high. Even so, the idea was sound and won a considerable amount of support from the local newspaper, the *San Jose Mercury*.

This was not the disaster it may have seemed at the time, however. Muybridge's failure in San Jose meant that he could give his full attention to a venture that would transform his life as dramatically as had his relationship with Flora. He was about to move into the big league.

SEVEN

MOVING PICTURES

On the 7 August 1877 in a letter (a copy of which I have before me) I suggested to you a plan for making a series of electro-photographs, automatically, by which the consecutive phases of a single stride could be successfully photographed. Being much interested in this subject, I offered to supply you with what copies of the results you required for your personal use if you would pay the actual expenses of obtaining them—omitting any payment for my time.

You accepted my proposition, and from a few days after the date of my letter, until the spring of 1881, or for more than three years, my time was devoted almost exclusively to superintending the construction of the apparatus or the execution of the work.

Letter from Muybridge to Stanford dated May 2, 1892

Leland Stanford probably never placed a bet on horses flying, but had the myth been true, Stanford would have risked less in his wager than he was about to invest in getting Muybridge to uncover the nature of motion, in particular the motion of horses. In total, the exercise would cost him over $40,000, around $800,000 in modern currency. But by then there was much more at stake than establishing that Occident really did fly.

Stanford was convinced that Muybridge's earlier work showed that the horse's four hooves did leave the ground simultaneously; now he wanted more, much more. As the entrepreneur threw himself into horse breeding he was determined to understand just how a horse moved. Where were the most important forces on its physique? Which muscles should be built up? How did weight distribution influence its capabilities? And in answering Stanford's questions, Muybridge was himself to move beyond simply freezing motion.

At the same time he was producing the San Francisco panoramas, Muybridge took a few more single photographs, first at the Sacramento track, then at Old Bay District Track in San Francisco, but it was only when Stanford acquired a property outside San Francisco for his horses that Muybridge was to go far beyond his original single-frame shots of Occident.

The property Stanford bought was the Palo Alto farm, which he acquired from "Lord" George Gordon, the man behind the South Park development. It is fitting that Muybridge should have made the first steps in shaping a key technology of the early twentieth century—motion pictures—at the same location that would later be the birthplace of essential elements of the leading technology of the late twentieth century—computing.

The Palo Alto farm later became the site of Stanford University. At a commercial offshoot of the university, the Palo Alto Research Center (or PARC), the computing company Xerox devised the technology that all our personal computers would eventually use. Mice, windows, menus, all the main elements of the graphical user interface were developed here. And just as Muybridge's work prefigured moving pictures but was not the specific technology they would use, so Xerox's efforts would inspire Apple and Microsoft without Xerox ever making any great commercial headway in this area.

Back in the 1870s, Palo Alto was largely open countryside. Stanford bought George Gordon's large county house, Mayfield Grange, and its surrounding 260 hectares (around 650 acres), then expanded the acreage by buying up surrounding land until he had amassed around 3,000 hectares (8,000 acres). Rather than continuing to call the place Mayfield, he picked on the romantic-sounding name Palo Alto (literally "high wood') that at the time applied only to a lone redwood tree that stood nearby.

Muybridge seems to have greeted the move to Palo Alto with mixed feelings. He wrote to Alfred Poett, Stanford's assistant: "Personally, I would as soon execute the work at Sacramento as at Palo Alto, were I assured that I would not be interfered with by people exercising horses, and others of the public." The one big advantage Palo Alto had, as far as Muybridge was concerned, was privacy.

It was at the Palo Alto farm that what had been a simple but clever technical development was to be transformed into the ancestor of a new medium. It is not clear whether the idea came from Muybridge himself or from Stanford—but the plan was not simply to capture a single image as a horse rode past, taking potluck on the position of the hooves, but also to take a whole series of shots, one after the other, allowing a study of the whole process of motion rather than a single slice of time frozen on a photographic plate.

It is tempting to imagine that the inspiration came from the San Francisco panorama, produced a few months earlier. For this Muybridge took a sequence of photographs, one after the other, moving the camera around through a fixed angle and then taking another. You could imagine Muybridge thinking, as he moved the camera on for the next shot, "If only I could do this fast enough I could unravel time, capturing movement in picture after picture after picture." The only problem was that it took him around 15 minutes to manage each exposure of the panorama and the subsequent repositioning of the camera. To succeed in doing the same thing as a horse went past would require the shot to be taken, the plate replaced, and the camera swiveled all in a fraction of a second. That clearly wasn't possible. Just as the early pioneers of factory automation would soon pull apart the processes of assembling products to give a small portion to each person in a chain, so Muybridge would have to pull apart the process of generating a series of photographs and assign each to a separate component of his photographic assembly line.

The comparison is apt. Where so far Muybridge's photography was effectively a cottage industry, it was to become the photography of the industrial revolution. With Stanford's money to back him, Muybridge ordered 12 top-quality stereoscopic cameras along with the ultimate in lenses of the time from Dallmeyer. While the cameras and other equipment were working through the order process, Muybridge and his assistants at the farm were far from idle. They built a remarkable structure in the exercise yard of the stable.

This time the photography was to be placed on a solid scientific

General View of Experimental Track, plate F from Muybridge's series *Attitudes of Animals in Motion* (1881).

basis. Carpenters and farmhands, more used to putting up barns and fencing, were directed by Muybridge in the construction of what would become known as the camera shed—though the name makes it sound less remarkable than it really was. On the south side of a mile-long track, already constructed alongside the farm for trotting training, was a long building, providing shelter for the cameras. It was like a huge carnival booth with a 50-foot-long counter. Inside the "booth" the cameras would be arranged, pointing out across the trotting track.

On the far side of the track, opposite the counter, a fence was built, 15 feet high, extending beyond the shed in both directions. (It was lucky that Stanford had no immediate neighbors—it must have been quite an eyesore.) The fence was draped with white sheeting painted with black vertical lines. Each of the tall, narrow boxes defined by the lines was numbered at the top. Even the track was not left in its natural state. It was sprinkled with lime in a powdered form to give a bright, reflective flooring. Though Muybridge was to use the most sensitive photographic chemicals of the day, he needed all the contrast he could get.

Contrast was the nightmare limitation of the early photographer. The photochemical reaction of negatives—bright parts of the image go black and dark parts of the image stay white—sounds simple enough, but the real world has subtle variations in contrast, often emphasized by differences in color. The human eye's perception of a scene, using its complex mechanism of rods and cones to separate out color and tonal information, is quite different from the crude measurement of light and dark offered by a photographic plate. To the early cameras such fine distinctions were lost. Unless a very long exposure was used, only the brightest and darkest parts of the picture could be clearly distinguished.

This difficulty with contrast is why the earliest photographs were all landscapes. Human faces, with their subtleties of tonal variation, could not be captured by the pioneering photographers (nor could the sitters remain still long enough for the lengthy exposures). When a portrait was first achieved, it required significant ingenuity on the part of the American who took it—a man whose other great achievement has eclipsed this one. It was Samuel Finley Breese Morse, the man who made the electric telegraph a reality and devised the code that bears his name.

Morse had been to Europe, visiting Paris in 1838, and there met Louis Daguerre. He was impressed by the early photographs he saw, and when he got home began to experiment with his own pictures. But, perhaps realizing the commercial possibilities of the portrait, Morse was determined to capture a human image (he had previously been both a portrait painter and sculptor). By 1839, Morse managed to take a photograph of his assistant, John Draper, though Draper suffered for Morse's art—he had to stay fixed in position for half an hour (for a number of years, portrait sitters would have their neck placed in a clamp to keep them still during the exposure), and his face was dusted with flour to increase the light reflected from his skin.

This need for increased contrast, particularly as exposures got shorter, would continue into the silent movie business, where skin makeup was initially a shocking white to ensure good rendering of the

Front of electrically operated shutters—plate C from Muybridge's series *Attitudes of Animals in Motion* (1881).

features. When Muybridge set up his track, he was aware that with the very quick shutter speeds he required—around 1/1000th of a second—he needed all the contrast he could get. The faster the shutter speed, the less light that made it through to the photographic plate. The less light, the more contrast that was needed to get an image to appear. As the horses were dark in color, he chose a white background, and used lime (slaked lime, calcium hydroxide, the main component of white-wash—a *San Francisco Chronicle* report of the time refers to it as "slack lime"—rather than the highly corrosive quick lime, calcium oxide), a brilliant white chalklike powder, to brighten up the track.

As well as constructing the location for the experiments, the Muybridge team had to work on improving the mechanics of the photographic process. How would the shutters of the cameras be triggered? For that matter, what would the shutters be? As we have already seen, most cameras at the time were little more than boxes for exposing a plate. There was no built-in shutter to let in the light for a carefully timed interval. The photographer took a picture by pulling off the lens cap, counting the seconds and replacing it, or by using a convenient shading device like a hat. Muybridge had not only to set out the cameras and trigger them he also had to provide a mechanism to enable the split-second exposures required.

The technology he was to employ was clever but simple. Bearing in mind that these were stereoscopic cameras with two side-by-side lenses each, he had to have either two shutters per camera or a wide shutter covering both lenses; he opted for the latter. Each of the 12 cameras needed a separate shutter, which would have to be capable of remote operation. It would not be possible to operate each camera directly by hand as he had in his earlier, single-shot experiments.

The mechanical shutters he constructed consisted of two horizontal wooden boards with a gap between them, like planks in a fence. These two boards were positioned in runners at either side, so they could move up and down together, fixed to each other with a gap slightly smaller than the camera lenses between the upper board and the lower one.

The boards were positioned in the runners so that the lower board began across the front of a pair of lenses, stopping light from getting into them. A latch held the shutter in place, while strong rubber bands pulled down on the boards. When the latch was released, the rubber bands, tensioned to a strength that could lift a hundred-pound weight, slammed the shutter boards downwards. As the shutter shot down, for a brief moment, the slot between the boards was in front of the stereoscopic lenses—the picture was exposed. Then the second board crashed into place and the lenses were covered once more.

Of course this approach was not perfect—early in the exposure only the top of the lens was receiving an image, while toward the end only the bottom was hit by light—but it did make it possible to take an extremely quick exposure. Before long the design was significantly improved by having a second, similar shutter in front of the first, traveling in the opposite direction. This stopped the "top first, bottom last" opening problem and cut down the exposure to the time when both the rising and falling slots were in front of the lens.

This shutter mechanism made it possible to take the exposure, but it also had to be triggered. With 12 cameras and a requirement for precision timing, it wasn't practical for the shutters to be activated directly by a hand pulling a catch on each shutter. Instead, Muybridge built in an electromagnet, so that the shutter could be released at the press of a button, on the completion of an electrical circuit.

This device might seem amazingly ahead of its time in an age of steam and horsepower, but Michael Faraday had discovered electro-magnetism, the principle behind the electric motor, as far back as 1821 at the Royal Institution in London. Although electrical devices still seemed modern and exciting in the 1870s, the electric telegraph had started sending messages under the Atlantic between the United States and Britain in 1873, and it was only one year after Muybridge's first Palo Alto experiment that the Siemens company built the world's first electric train at Bush Mills in Northern Ireland. (The first public electric railway came into use only four years later, in 1883, on the Brighton seafront in England; it is still running today.) This was very much the time of electrical firsts.

Not long before, electricity had been constrained to the realm of cranks and parlor tricks. Great healing powers had been claimed for electricity—after the Italian Luigi Galvani had demonstrated in 1771 that electricity could make a detached frog's leg twitch, it had been seen a sort of vital force of life, a role that Mary Godwin (soon to be Mary Shelley) would make good use of in her 1816 gothic novel, *Frankenstein*. Faraday had brought electricity into the respectable hands of science and engineering and now the applications were flourishing everywhere.

As far as Muybridge's work was concerned, the important discovery was not so much Faraday's synthesis of electricity and magnetism that led to the development of the electrical motor, as the earlier work on electromagnetism that Faraday researched. In 1821, Faraday, then little more than a lab assistant at the Royal Institution, had been given the task of pulling together what was known about electricity and magnetism. Always a hands-on person, rather than simply a reviewer of information from other papers and books, Faraday repeated all the experiments he could find describing electrical work around magnets.

Most of what was then known had been discovered during the last year. Danish experimenter Hans Christian Oersted had found that a wire carrying an electrical current produced an effect like a magnet, while two French scientists made a more practical proposition: André-Marie Ampère had produced an effect more directly like a magnet with a coil of wire, while Dominique François Jean Arago had found that an iron or steel bar could be magnetized by putting it inside the helix of a current-carrying wire. From the information Faraday collected, it only took one further development, by the English electrical experimenter William Sturgeon in 1825, who discovered that leaving the metal bar inside the coil made the whole setup act as a much stronger, more practical magnet, which made Muybridge's electromagnetic shutters possible.

Muybridge's electromagnetic shutter combined subtle and surprisingly crude aspects of construction. Just like the mechanical shutters, a pair of horizontal wooden slats with a gap in between to provide the exposure were held up at the top of a wooden frame by a latch, resisting the pull of a pair of rubber bands. This latch passed through the frame to an arrangement on the outside that looked like an early Morse key. However, instead of the key being pressed by a human hand, an electromagnet attracted the key toward the frame. As the key was balanced on a fulcrum, the opposite end was pulled out of the frame and released the latch.

Exactly how much of the design was done by Muybridge himself and how much was done by assistants has been the subject of much dispute and was later to confuse many historians. Muybridge constructed a rough model of his high-speed shutter, which he showed to Stanford. It seemed to the governor that it was in need of some professional engineering enhancement, and he sent Muybridge to see his chief engineer at the Central Pacific Railroad, Samuel Montague. Montague delegated the job to one of his best engineers, Arthur Brown, who helped refine Muybridge's design.

At that stage the design appears to have been mechanical. One of Brown's actions was to bring in a junior engineer John D. Isaacs, who had experience with electrical devices, to help with the electromagnetic version. We only have the word of those involved as to who originally thought of using electricity in the shutter. Isaacs would later claim it was his idea—which seems unlikely, as he was called in specifically for his electrical expertise, implying that someone had already thought of this approach. What's more, Muybridge had visited the Electrical Construction House on San Francisco's Sutter Street in September 1877 with Stanford's assistant Poett and paid the San Francisco Telegraph Company $130 in November 1877 for the supply of "Electrical Photo Apparatus."

This commercial electrical activity seems to have taken place before Isaacs was involved in the project. This makes it more likely that Muybridge or Brown had the idea of using an electromagnetic release, though a small handful of writers have bizarrely used this incident to "prove" that Isaacs and not Muybridge was the father of the moving picture.

When the cameras and lenses finally arrived, they were positioned inside the shed at a spacing of 21 inches apart, just as the lines on the white drape on the fence across the track were 21 inches apart. Why 21 inches was chosen is not clear, though it may have been to accommodate the 20-foot "stride" of the horse with the 12 cameras available. There is some confusion over the spacing, as in a talk about the experiments given in 1882, in which Muybridge referred to the lines on the drape as being 12 inches apart, while the contemporary newspaper report (see below) has them 2 feet apart. However, the annotation of the photographs clearly says 21 inches, which fits with the scale of the photographic image. It looks like the newspaper was rounding, while the report of Muybridge's talk assumed that the "21" was a misread 12.

Now all that Muybridge had to do was to trigger the shutters as the horses passed by. Those electromagnetic shutters not only made it possible to handle all 12 cameras, but also provided a mechanism for the horses to take their own photographs in a precise fashion, initiated by their progress along the track.

When the horse was pulling a carriage—usually the lightly built trotting carriage called a sulky (so called because it was built for one person only, the implication being that someone who wanted to ride in a carriage like this was sulky)—the wheels of the cart were rimmed in metal. Along the track a pair of wires ran to each camera. When the sulky's wheels pressed down on the wires it completed the circuit between the two, sending an electrical pulse to the magnet that released the shutter.

The *Alta California* reported all this in slightly breathless but factual style:

> Mr. Muybridge has now received his instructions [from Stanford] and will commence his work as soon as he can receive the needful lenses from London, and can have some machinery made here. "Occident" moves 20 feet at one stride, and Mr. Muybridge will have a dozen photographic cameras

placed at intervals of 2 feet, making a total distance of 24 feet, a little more than a full stride. The shutters of these cameras will be opened and shut by electricity as the horse passes in front of each, the time of exposure being as before not more than the thousandth part of a second. The twelve pictures will be taken within two-thirds of a second, the time required for traveling 24 feet at a speed of 2.37 [i.e. a mile in 2 minutes 37 seconds]. Each picture will be taken by a double lens, so as to be adapted for the stereoscope, and will thus furnish the most conclusive proof to connoisseurs that it is faithfully taken by photography and not materially changed by retouching.

Tripping the shutters electrically in this way was fine, but Muybridge had no intention of limiting his studies to carriage horses. He wanted to be able to photograph freely running animals, which would require a different trigger mechanism. This was something he already seems to have tried out in his experiments while at the Old Bay District Track in San Francisco while he was waiting for construction to be finished at Palo Alto. His assistant at the time, a boy of around 16 called Sherman Blake, would write in his old age that "the horse sprang the instantaneous shutter, by means of the thread across the track."

This was the simple mechanical alternative Muybridge used to his electrical shutter releases. Threads were stretched across the track at the horse's breast height. As the horse struck each thread, it pulled on a wire, making a connection to trip the shutter. The thread then broke. On the whole this technique worked, but it was significantly less reliable than the direct electrical trigger, and there was a danger of a horse being spooked by repeatedly hitting the threads. In fact a contemporary report of the first use on a racing mare called Sally Gardner makes this clear:

> When the mare broke the eighth or ninth thread she became aware of something across her breast, and gave a wild bound into the air, breaking the saddle girth as she left the ground. This gave a curious picture of the mare with her legs wildly spread and the broken girth swinging in the air just as it is separating.

It had taken all winter and spring to get the apparatus correctly set up. It was not until June 11, 1878, that Muybridge was ready to take any photographs. He took the risky step of arranging to take the first shots with the press present. It wasn't just a matter of self-aggrandizement; he was concerned that otherwise he would not be believed.

Bear in mind that retouching was a universal process at the time, and that his previous photographs had certainly gone through this process. To Muybridge this was just a sensible part of making a photograph suitable for human viewing, but a more suspicious mind might have thought otherwise. Retouching sounds harmless, but was effectively turning a pure photographic record into a hybrid of a photograph and a painting—and that meant that the image could be made to show whatever the photographer liked. Oddly, we are now in exactly the same position with digital images, which cannot be relied on as evidence as they are so easy to modify.

To make matters worse, at least one newspaper would simply not believe the split-second timing of the photographs. After all, anyone could pose a horse and cart in front of a camera and say that the shot was taken at speed. There were those who were just as doubtful of Muybridge's claims to have captured high-speed trotting as later skeptics would be of the suggestion that the photographs and film from *Apollo 11* really proved that man had landed on the Moon.

It was, perhaps, no coincidence that the critical newspaper that doubted Muybridge was the *San Francisco Post,* the paper that had employed Harry Larkyns. Despite the fact that Larkyns had been dismissed for misleading his employers, the *Post* remained surprisingly loyal to their ex-correspondent. Nine months earlier in September 1877, when Muybridge had taken the last of his individual photographs at the Sacramento course and had claimed to have frozen Occident flashing past at the speed of a mile in 2 minutes 27 seconds (around $24^{1}/_{2}$ miles, or 39 kilometers, an hour), an anonymous journalist from the *Post,* writing under the bizarre title "Bohemian Bubbles from a Rambling Writer," was scathing. He describes Muybridge's claim with amusement:

> It is a triumph of which we would say in the Celtic tongue, "Moryah." I will not swear that I have spelled the word correctly, but I have given the sound. The word has the same strength as humbug multiplied say ten thousand times.

This term "moryah," which James Joyce was fond of, spelling it "moya," is described in the Oxford English Dictionary as conveying "the same sense as the English saying, 'You may tell that to the Horse Marines.'"

But why was the writer so unimpressed? He could hardly admit to being biased against Muybridge because of the Larkyns affair. Instead he went on to justify his opinion by decomposing the photograph mercilessly.

> Let us look at this triumph of an art seemingly in its infancy: The driver, Mr. Tennant—I presume it is Mr. Tennant, though I do not enjoy the gentleman's acquaintance—is not driving a horse; he is sitting for his photograph. He is stiff, un-natural; he does not encourage his horse; he would lean forward were he driving at the rate of 36 feet per second; he would be alive with movement and the "hie yar" would, as it were, ring in our ears.

The Rambling Writer goes on to sarcastically comment on Tennant's driving:

> If Mr. Tennant drives a trotting horse in this fashion, when I come in for that fortune—which, by the by, the rightful owners are keeping me out of—he will never drive one of mine. Decidedly, Mr. Tennant, you were not driving Occident at the rate of 36 feet per second when you sat for that photo.

As if it's not enough to knock poor Tennant, he goes on to have a poke at Leland Stanford. After making ludicrously detailed measurements of leg positions from what was, after all, a fairly poor snapshot he goes on:

> And here's the rub. Either that camera did lie, or Stanford has got the most extraordinary horse in the world. The "woolly horse" of Barnum was not a circumstance to it and he can make more money by exhibiting it than be trotting it.

Rambling Writer tries to soften the blow a little by commenting, "Seriously, I may appear to be more of a cynic than a critic, but this is the way the photograph of Occident by Muybridge strikes me." Even so, the result is painfully insulting. Faced with such outright mockery from at least one member of the press, it seems that Muybridge and Stanford were determined to leave nothing to the imagination in their new, much more dramatic experiments. At Stanford's request, his sporting friends were joined by a group of newspaper reporters to monitor exactly what happened. They saw first a sulky, pulled by a horse called Abe Edgington and driven by Charles Marvin, the chief trainer at Stanford's stable, and then the mare Sally Gardner run the gauntlet of the cameras. The sulky did a timed 2 minutes and 20 sec-

Sallie Gardner galloping (2), 1878—phase 2 from a series of eight lantern slides.

onds on the mile track (25.7 miles, or 41 kilometers per hour) while Sally Gardner is described as "rushing down the track like a whirlwind."

The plates were taken straight from the cameras to a developing room, right on the end of the camera shed, and processed while the gathered reporters waited. Within 20 minutes the negatives were on view, to the amazement of the reporters, who declared it a "brilliant success":

> It is difficult to say to whom we should award the greater praise, Governor Stanford, for the inception of an idea so original, and for the liberality with which he has supplied the funds for such a costly experiment, or to Muybridge for the energy, genius and devotion with which he has pursued his experiments, and so successfully overcome all the scientific, chemical and mechanical difficulties encountered in labors which had no precedent, and which have so happily culminated in such a wonderful result. We hope that he will reap the benefit to which his genius and success so clearly entitle him.

It is significant that it is Muybridge's genius and success that are celebrated, as attempts would later be made to rewrite history and minimize Muybridge's contribution. Another contemporary report was equally enthusiastic, and similarly made a clear division between the basic idea (and cash), attributed to Stanford, and the practical genius to make it happen, which was all due to Muybridge:

> Even the thread-like tip of Mr. Marvin's whip was plainly seen in each negative, and the horse was exactly pictured. . . . Mr. Muybridge, a photographer of genius and an artist of rare skill, was the operator. He is, practically, the inventor of the process, although the design or idea was suggested by Governor Stanford, who has unstintedly [sic] supplied the means to perfect the apparatus.

What Muybridge could not have anticipated was how fortuitous would be the breakage of Sally Gardner's girth when the horse was frightened by passing through the threads. In principle, photographs of a horse simply passing the camera could have been faked in some way, but when the photographs were developed, and they showed the detail of the accident that the reporters had just witnessed, this was undisputable proof. As the *San Francisco Morning Call* remarked:

> The girth breaks and she bounds as she leaves the blanched ground. This gives a queer picture, far more strange than the other. All previous notions are dispelled, and what was thought to be the more simple is far more complex. The fidelity is equally as remarkable; the broken girth is portrayed the instant it separates, and every skeptic is a firm believer.

Muybridge promptly patented both the combination of the electromagnetic shutter with the distance-marked back fence and a second shutter operated by breaking a thread that used a mechanical, spring-tensioned mechanism (surprisingly, the mechanical shutter is described as an improvement on the electromagnetic version). It doesn't help Isaacs's claim to have invented the shutter that he never disputed these patents, though he later said that he had confronted Muybridge about it. According to Isaacs, he met Muybridge in Morse's gallery, where Muybridge was triumphantly showing off his scrapbook of cuttings. Isaacs complained that he couldn't "see much of Isaacs in it." According to Isaacs, Muybridge then took umbrage, remarking stiffly, "What did you have to do with it, sir? You were only employed as a draughtsman for me."

Isaacs, feeling that his role was much more significant, was tempted to sue Muybridge, and only failed to do so because his boss, Arthur

Brown, persuaded him that it was all Stanford's business, whoever did the inventing, and so the matter should not be taken to court. His claim would not surface again for many years—for now, as far as the newspapers were concerned there were only two names involved—Stanford and Muybridge.

The demonstration on the 11th of June was not to be a one-off experiment for the benefit of the reporters, though. Muybridge went on, refining the experiments, trying different shutter mechanisms and chemical mixes, through the next year. He was often accompanied by newspapermen (cynically, it may have been seen as an easy story when things were quiet). There was no further attempt by the *Post* to claim that Muybridge was faking the results, but there was one letter writer who refused to follow the party line praising the remarkable work at Palo Alto—and strangely, it was a man who had been one of Muybridge's closest friends.

∞∞

Muybridge seems never to have forgiven William Rulofson for his testimony at the murder trial. Rulofson was just too convincing in putting across the message that Muybridge was unhinged. But equally, it appears that Rulofson was unhappy with his treatment by Muybridge. After all, he had only been acting at the request of Muybridge's attorney, and had been trying to prevent his friend from being hanged.

Given the sheer bile of a letter Rulofson had published in the August 1878 edition of the popular photographic magazine *The Philadelphia Photographer,* it seems likely that by the time a few years had passed, Rulofson's initial bewilderment had turned into bitter resentment:

> None know better than yourself that the country is full of photographic quacks vending their nostrums, deceiving the credulous and defrauding the innocent. California is noted for its "largest pumpkins," "finest climate," and "most phenomenal horse" of the world. So also it has a photographer! [*sic*] the dexterity of whose "forefinger" invokes the aid of electricity in exposing his plate—a succession of plates—so as to photograph each particular respiration of the horse. The result is, a number of diminutive silhouettes of the animal on and against a white background or wall; all these in the particular position it pleased him to assume, as the wheels of his chariot open and close the circuits.

All this is new and wonderful. How could it be otherwise, emanating as it does from this land of miracles? Photographically speaking, it is "bosh"; but then it amused the "boys," and shows that a horse trots part of the time and "flies" the rest, a fact of "utmost scientific importance." Bosh again.

Hardly the contribution of a friend and supporter. It's possible that Rulofson had only seen the earlier, separate photographs, which were little more than diminutive silhouettes, but this was a totally inappropriate description for the new experiments. A few months later (as it happens, after Rulofson's death), the editor of the magazine felt it necessary to respond at some length. He begins by working through the photographic sequence, as he puts it, "going into them":

> The succeeding positions in No. 1 we can get through without much excitement. There is a great deal more apparent "go" in No. 2, and yet more in No. 3, but when we come to Nos. 4 and 5, we get the "poetry of motion" twelve times intensified in every nerve. We imbibe all the energy of the horse. We stretch our imagination to its maximum and are forced to cry "stop," Mr. Muybridge, you have caught more motion in your photographs than any previous camera dreamed of. They are truly wonderful, and we congratulate you.

He goes on to counter Rulofson's remarks:

> Surely Mr. Rulofson could not have given much thought to the subject when he called these pictures "silhouettes" only, and the claims made for them as "bosh." Mr. Muybridge deserves great credit, and has gained great notoriety for what he has done, and we shall try to induce him sometime to tell us more about it.

In fact, it seems more likely that Rulofson had given the subject far too much thought—at least, too much thought to the subject of Muybridge wronging him—and had been hoping to suppress the growing wave of enthusiasm for Muybridge and his work. His attempt was as effective as pouring a single bucket of water on a forest fire.

One day in November 1878, between the publication of his original letter and the editor's reply, Rulofson climbed up onto the high roof of his gallery to check progress on a skylight that was being fitted to improve light levels in the studio. He slipped and fell to his death. (There is no evidence to support the allegation of the writer Ambrose Bierce that Rulofson committed suicide. Bierce and Rulofson were in dispute over the authorship of *The Dance of Death*, a strange book published under the pseudonym William Herman that claimed [as a straight-faced joke] that the waltz was "an open and shameless gratification of sexual desire and a cooler of burning lust." This disagreement may well have triggered Bierce's typically over-the-top comment.)

It wasn't enough, though, for Muybridge to have won the support of the press. He had always intended the photographs to be the basis for public lectures and exhibitions. The very first such event, in San Francisco on July 6, 1878, was poorly attended, but word of mouth and the newspapers spread the news that it was an evening well worth attending, and a later event that same month drew a good crowd.

It is interesting that the *San Francisco Examiner* reported that the lecture was "crowded with an intelligent and fashionable audience, and which was feminine by a pleasant majority." It would be tempting to assume that Muybridge's audience was largely composed of men interested in the racing implications of the photographs, the racing fraternity being largely male at the time, but it seems instead that the artistic merits and sheer fascination of his work (combined with the showing of photographs from Yosemite and Central America) brought in a much more cosmopolitan crowd.

Through 1879, Muybridge improved on his experiments. The number of cameras was increased from 12 to 24, doubling the amount of movement that could be analyzed. And at the instigation of Stanford's wife (Muybridge later wrote to Stanford that it was as a result of "Mrs. Stanford's . . . desire to extend the investigation") he began to move away from horses. During the summer a whole string of animals—cows, deer, dogs, goats, hogs, and oxen—were sent down the Palo Alto track, which by now must have been looking a little the worse for wear after the passage of such a motley crew of animals that left behind everything from footprints to manure.

While some of the animals—the dogs, for example—could be relied on to make a reasonably steady progress through the threads, many of them were significantly less cooperative. Yet another means of triggering the shutters had to be found. For these animals Muybridge used a clockwork mechanism that triggered shutter after shutter in a regular action, like the timed initiation of notes in a music box. By this time there were plenty of mechanisms for achieving a slow, steady rotation driven by a clockwork motor. All Muybridge had to do was have the same type of rotating cam striking a series of electrical connections that would provide the trigger for the appropriate relay to take the picture.

As the device rotated, one after another of the cameras' shutters were launched into action. Muybridge would patent this device in 1883 as a "method and apparatus for photographing changing or moving objects." His new timer meant that the sequence of shots became a time-based record, rather than the space-based approach where the passing animal, however slow or fast it moved, triggered the shutters at regular intervals of distance as it moved along. The result was a much more accurate portrayal of the sequence of motion, and would be the basis of most of Muybridge's future photography. This first timer was also like a music box in having a horizontal rotating cylinder with a series of prongs along it that touched a line of contacts; in later years he would use a more advanced model with a circular set of contacts that was much more flexible.

By August, the Noah's Ark collection of creatures trooping in front of the lenses had inspired another idea. If it could work for animals, it could work for people. After all, even though Muybridge was no portrait photographer, there was no subject more fascinating to audiences than the human being, and for artists the ability to study aspects of the human frame in different stages of motion could greatly enhance their ability to realistically portray lifelike scenes.

It may well have been that this was at the insistence of Stanford. The governor had links with a San Francisco athletic club called the Olympic, and arranged for Muybridge to take a number of classic group photos of the three men who turned up as subjects, which Stanford then sent on to Europe to have copied as oil paintings to be hung in the club. But these poses were only taken after the men had spent a grueling six hours wearing "only brief trunks" undergoing a range of athletic pursuits from boxing to jumping. For Muybridge these sequences would be the start of something much bigger than the horse photographs: his great study of human motion.

For now, though, Muybridge continued to take photographs at Palo Alto until late 1879, and experiments went on sporadically for another two years, though he was not in town often enough to add substantially to his portfolio. Even those who had argued with Stanford over the "flying" horse were now totally sold on the effectiveness of the Muybridge photographs. George Wilkes of the New York *Spirit of the Times*, the man most likely to have bet against Stanford had there been any bet, was magnanimous in print, if rather sidestepping his own earlier position:

> It has long been a vexed question as to whether a trotting horse was ever clear of the ground while in motion, and some lights of the turf have stubbornly argued the negative of the proposition. These gentlemen must now, as gracefully as possible, lay aside opinions held for years.

The acclaim of the photographs and Muybridge himself stretched countrywide and across the Atlantic to be picked up with fascination in Europe. The French scientific magazine *La Nature*, which had covered the photographs in the previous edition, carried an enthusiastic letter from Professor Etienne Jules Marey in the December 28, 1878, edition:

> I am lost in admiration over the instantaneous photographs of Mr. Muybridge, which you published in the last issue of *La Nature*. Can you put me in correspondence with the author? I want to beg his aid and support to solve certain physiological problems so difficult to solve by other methods.

The French professor, who had earlier attempted to monitor animal motion mechanically, conducted an effusive correspondence with Muybridge, both in the press and directly, which emphasized that Muybridge's work was not just a wonder for the common herd but of scientific importance, too.

Interestingly, the next edition of *La Nature* carried an article about a totally different invention of Muybridge's, showing that even at the height of his photographic work he was not above turning his mind to other things. Under the heading *Horloge Pneumatique*, the article tells how "M. Edward J. Muybridge de San Francisco," already known to their readers, is the inventor of this ingenious instrument. At its heart is a perfectly normal clock, driven by a weight and regulated by a pendulum. But as the escapement shifts the clock hand tick by tick, it also raises and lowers a bulb-shape object that appears to be made of glass (though this might have been just to make the diagram clearer). The bottom of the bulb, which is open-ended, descends into a liquid, so the air in the bulb is compressed, runs along a fine tube, and there activates the escapement of another clock, so the two ("or more!" the article triumphantly suggests) clocks can be kept exactly in time with each other.

The principles behind the clock are sound enough, if easily eclipsed by the sort of mechanism Muybridge had already used in his electromagnetic shutters. Whether the clock was ever built is uncertain—it probably wasn't—but it was a novel design and typical of Muybridge. He obviously spent some time on this device, as the drawings in the magazine, which show both a master and slave clock, were professionally made for him by the firm of Van Vleck, Engineers.

However, the clock was merely a brief distraction. For a man with Muybridge's grasp of the economic benefits of fame, the exposure he was getting for his horse photographs was a situation he was not going to waste. It was time to make the move from backroom boy to celebrity.

EIGHT

TAKING EUROPE BY STORM

Muybridge had never been a shy man. This was, after all, the boy whose cousin had considered him "rather mischievous" and who had been prepared to launch himself onto a new and largely unknown continent with only his wits to live on. Yet until now his interface with the public had been indirect. It was his photographs, rather than Muybridge himself, that had been the link with his audience, a fact that was emphasized by his early pseudonym of Helios. Yet like today, public acclaim was accompanied by the demand to be seen. With no TV, this meant public appearances—and Muybridge found, quite possibly to his own surprise, that he enjoyed making them.

From around the middle of 1878 and increasingly as the Palo Alto experiments went on, he began giving public lectures, illustrated by his images in lantern-slide form—large-scale transparencies, projected onto a screen. Not only did this reflect his increasing celebrity, it was also a useful source of income. Although Stanford had paid a huge bill for the photographic work—well over $40,000 by the end of 1879—Muybridge still had his own expenses to cover and the hope of making some profit on the side.

The main visual approach he used at the time was a side-by-side comparison. By having two lantern-slide projectors, he could use one to show a condensed view of the full set of images of a horse's motion, while the other projector was used to pick out detail in slide after slide. He could, and did, also use this setup to mock the attempts of painters to portray horses in motion, contrasting the stiff, artificial poses in paintings of different ages with a similar but much more natural photograph, a comparison that regularly drew laughter from his audience.

For a while these lectures were very popular in San Francisco and the surrounding cities, but the novelty palled. By now, though, Muybridge had other possibilities in mind. Out on the track, as we have seen, he had moved beyond photographing horses alone to begin a wider study of animals in motion. For his lectures he now conceived of a much bolder step forward. Not only would he show his moments of frozen time one by one but also string them together to re-create living motion on the screen.

The stimulus for this idea may have come from the editor of the leading science magazine, *Scientific American*. An 1878 issue had reproduced a full sequence of images of a horse in motion (photographs still could not be reproduced in normal print at the time, so these were engravings taken from the originals). The editor commented in passing that it would be possible to cut out the pictures, paste them onto a strip of card, and re-animate them using a zoetrope.

The zoetrope was a toy of the sort that amused children but also found its way into the drawing rooms of adults. It had been around for about 20 years, and was losing popularity by the mid-1870s, but this suggestion from the *Scientific American* editor would help give it a new lease on life. It made use of two distinct features of the eye and brain working together. The first feature exploits the brain's tendency to suppress very short blank periods of input—the second is the way that the brain treats a series of still images that involve small, sequential changes as if it were seeing true movement. For a long time these effects were explained as "persistence of vision." This idea, first put forward around the same time as the emergence of the movie industry, depended on the idea that some sort of after-image remained in the brain long enough to overcome the blank gap while the picture was changing to the next one, and that the two slightly different images then merged together to form the effect of motion.

Such was the early enthusiasm for the concept of persistence of vision that it is still widely assumed to be the true explanation for the two movie effects today. Unfortunately, more recent research makes it clear that after-images don't form until around 50 milliseconds after the image has ceased to be projected, which isn't quick enough to bridge the gap between frames. Practical experience from the early days of cinematography showed that you had to change the pictures around 50 times a second to fool the eye. Early silent movies were shot at around 16 frames per second, with each frame shown three times, while sound movies run at 24 frames per second, showing each frame twice. The images are on screen for too short a time for persistence to account for the lack of visible flicker. And persistence of vision was never an adequate explanation for the second effect, apparent motion, as persistence would result in multiple images building on top of each other, not in the appearance of movement.

What is often described as persistence of vision is more accurately a reflection of the brain's ability to interpolate and substitute what it thinks is the right thing to see for the actual visual signal it is receiving from the optic nerves. The concept of persistence of vision relied on an outdated idea that the eye was like a camera obscura, projecting images onto the "screen" of the brain. In fact the brain contains a range of different visual sensory "modules" dealing with requirements like motion detection, object and pattern recognition, detail selection, and so forth. (These modules are conceptual rather than physical; they don't uniquely occupy a single set of brain cells.)

These different modules don't handle a single picture, but rather many different elements. The retina of the eye contains around 130 million light-sensitive receptors. When a photon penetrates to the back of the retina (the photoreceptors are back-to-front with the sensitive part at the rear, a clumsy arrangement that may well be an accident of evolution), it triggers a photochemical reaction. This reaction sends a signal back toward the surface of the retina, where input from different receptors is combined before feeding the information through the optic nerve to the brain. This nerve has a lot fewer nerve fibers than there are receptors in the eye, so the signal has already been processed before reaching the brain.

The combined image we "see" is much more an illusion than it appears, being a reaction to these complex inputs and a combination of the response of the brain modules that cope with motion, pattern, detail, and so forth. The suppression of the flicker between frames of a movie and the merging of still pictures into motion is not due to simple persistence; it is a side effect of the way the various complex systems involved in processing the optical data work together.

A similar example of the brain "cheating" and ignoring the true input it receives to achieve a useful outcome is the apparently steady view we have of the world. In reality our eyes spend a lot of time darting about and are very rarely still. Whether we are looking at someone else's face or the text in a book, our eyes undertake tiny, fluttering motions called "saccades." From the differences in outlook generated by these movements, the detail system in the brain can build up a much more sophisticated image than we could manage if everything were taken in via a static, camera-like gaze. Saccades are very quick—in fact, they are the fastest of all external motions of parts of the body, sweeping through an angle of 10 degrees in as little as 1/100th of a second. We can't cope with such a disrupted and blurred outlook; therefore, much of the information is simply edited out by our brain—used computationally, but not presented in our apparent vision. This misrepresentation of input is much closer to what is happening when we watch a movie than is the concept of persistence of vision.

The brain's complex handling of moving pictures causes the resultant display to seem to move naturally if an image can be quickly replaced while hiding the intermediate motion required to get the new picture into position. The simplest way of getting pictures replaced quickly enough is a flick book—a sequence of drawings on the pages of a small book that gradually change position, which can then be animated by flexing the book with the thumb on the edge of the pages and quickly riffling through.

Toys making use of this effect started to become common in the 1820s. The earliest and simplest example is the thaumatrope, which sounds much grander than it truly is. Commercially introduced by British scientist John Paris (though many think it was originated by the astronomer John Herschel), it consists simply of a circular disc with an image on each side. The disc is spun on a string, merging the images from the two sides as they are quickly and repeatedly brought in front of the eye, for instance, placing a canary pictured on one side in a cage on the other.

The thaumatrope didn't so much give an impression of motion as magically combine two different images. But the next in line of these strangely named contraptions did try to mechanize the flick book effect. The phenakistiscope was produced in 1830 by the Belgian Joseph Plateau (and pretty well simultaneously under a different name in Austria by Simon Stampfer). This device consisted of a flat disc with a series of images around the outer rim. Between each image was a slot. The disc was mounted on a rod, facing away from the viewer, who looked through the slots at a mirror behind, seeing each of the reflected images through a slot as it passed by.

The zoetrope, invented in the United Kingdom by William Horner in 1834, was a more practical version of the same effect. The equipment is like a deep, circular cake tin mounted on a swivel in the center of the bottom of the tin so the cylinder can be freely rotated. A series of slots are cut in the side of the tin at regular intervals. When you look horizontally at the tin, focusing through the slots, and rotate it, your eye is presented with a series of glimpses of the opposite interior of the tin, without seeing the intermediate positions that are screened by the solid parts of the tin.

Now fix a series of pictures on a strip around the interior of the tin, one for each slot. If each picture is slightly different from the rest, the eye will merge them together as the tin rotates. The result is a crude moving picture. Horner never really made much income from the daedelum, as he called his device; it was only 30 years later, when William F. Lincoln relaunched it as a commercial product in the United States as the zoetrope, that it took off in a big way.

Scientific American suggested that those drawings around the inner rim of the zoetrope be replaced with Muybridge's photographs and the result would be almost as if you were watching an actual horse in motion. A number of commercial ventures were to provide zoetrope strips of the Muybridge horse sequences, often with little or no profit getting back to the photographer.

But the zoetrope had severe limitations. The size of the device limited both the quality of the images and the number of people who could look through the slots at one time (each viewer, of course, saw a different part of the sequence at any one time). Because most of the time you are looking at the blank outside of the tin with only a brief glimpse through the slot, the resultant image is very dim. And the mechanism also means that the animation is jerky, as the images are not replaced quickly enough for the brain to conceal the flicker.

After a number of trials with a limited experimental rig, Muybridge took a giant leap forward from the zoetrope to a device that made it practical to show moving pictures bright enough, large enough, and smoothly enough to entertain and amaze a large audience. He called his new machine the "zoögyroscope." (That dieresis on the third letter is to indicate that the two oh's in the device's name are to be pronounced separately, something like zoh-oh-gyroscope.)

Perhaps to avoid confusion with the gyroscope, which had been named by the French scientist Foucault in the 1850s and became a popular parlor entertainment, Muybridge was to quickly rename his moving picture projector the "zoöpraxiscope." Instead of the glass slide normally found between the lamp housing and the lens of a magic lantern, Muybridge's projector had a pair of flat, circular rotating discs, one carrying a series of images around its circumference, the other opaque with regular clear gaps. This second disc, made of sheet metal with slots cut in it, rotated in the opposite direction to the picture disc, acting as a form of shutter, illuminating the image only when it was lined up with the projection system. The picture disc, typically 12

The zoöpraxiscope.

inches in diameter, would feature anything between 12 and 24 images, painstakingly arranged around the edge of the disc.

Initially, it seems that the picture disc showed only silhouettes, painted from the Muybridge sequences, but a photograph of a later disc shows a considerable amount of detail on each picture. Even so, these may have been painted onto a photograph, or from the photograph, rather than being a pure photographic original. Many of the discs Muybridge used were hand drawn rather than photographic, originally to overcome the poor contrast of the photographs and also because of a peculiarity of the zoöpraxiscope; to look normal on the screen the images had to be unnaturally elongated. It has been sometimes suggested that this distortion of the images shows that they were never photographic, but in Muybridge's preface to his book *Animals in Motion* he makes it clear that this distortion was intentionally produced by optical means when printing photographs for the zoöpraxiscope:

To correct the apparent vertical extension of the animals when seen through the narrow openings of the metal disc on its revolution in such close proximity to, and in the reverse direction of the glass disc, the photographs on the latter, after numerous experiments, were ultimately prepared as follows.

A flexible positive [photograph] was conically bent inwards, and inclined at the necessary angle from the lens of the copying camera to ensure the required horizontal elongation of the animal while the straight line of ground corresponded with the curvature of the intended ground-line of the glass disc, towards the periphery of which the feet of the animal were always pointed.

A negative was then made of this phase, and negatives of the other phases in the same manner. All the negatives required for that particular subject were then consecutively arranged, equidistantly, in a circle, on a large sheet of glass.

The elongation was required because of an effect that had already been observed when drawing pictures for the zoetrope: the image seen through the slot seemed to be squashed up, because the slot panned across the image, compressing its appearance in time, so it was only by using an elongated version that a natural-looking picture was viewed. Similarly the zoöpraxiscope suffered from the slot moving in the opposite direction to the image, causing a foreshortening and requiring a lengthened picture to look normal. This was a kind of inversion of the anamorphic process used in filming modern motion pictures.

Anamorphic movies were introduced to battle a commercial enemy. When, in the 1950s, TV began to challenge the movies, Hollywood panicked. With movie theater attendance in decline, it seemed natural to search for a way to make the movie experience different from crowding around a little box in the home. Movie sound quality was improved, and there was a brief dabbling with three-dimensional effects (never entirely satisfactory as the audience had to wear cheap eyeglasses, and the movies always seemed to give the viewer a headache). The solution that was to transform the appearance of the movies was widescreen.

Originally movies, like TV, were photographed using images that were in the ratio of 4:3, which is more easily compared to other sizes by thinking of it as 1.33:1. The picture was 1.33 times as wide as it was high. (Technically movies were ever so slightly wider at around 1.37:1, but this difference from the TV ratio is negligible.) The reasoning behind these values is submerged in the myth of Thomas Edison. Early still photographs varied randomly in format between square (1:1) and around 1.5:1. Edison decided to use Eastman's new 70-millimeter-wide celluloid film for his moving pictures but thought the stock was too big and so wasted money. When Edison's assistant, William Dickson, asked what the moving picture images should be like, Edison apparently answered "About like this," holding his fingers in a shape Dickson interpreted as 1.37:1.

Just why Edison came up with this shape has never been explained. Some have suggested it was inspired by the classical "perfect" ratio of the golden section, which is near 1.6:1; however, such classical allusion does not fit well with what is known of Edison's character (and lack of formal education). It's much more likely that he was simply indicating that he wanted the 70-millimeter stock cut in half down its length, the most natural way to split it, and this was all the shape was intended to indicate. This ratio was adopted as standard by the Society of Motion Picture Engineers in 1917.

Unfortunately a 1.37:1 ratio is a poor representation of the natural view produced by our eyes, which is significantly wider. (Try holding your hands together, one on top of the other, horizontally in the center of your field of vision. Move the top one up and the bottom down, until they get to the edge of your vision. Then do the same with your hands held vertically together and separate them to left and right. There is a much greater extent of vision horizontally.)

So to get a more realistic screen image that emphasized the difference between movies and TV, Hollywood began to shoot in a "letterbox" format that varies between around 1.66:1 (close to 1.77:1 or 16:9, which is the format used by modern widescreen TVs, originally developed as a compromise ratio for electronic storage of widescreen) and as much as 2.35:1 (the modern Cinemascope and Panavision). Reflecting the interest in 3-D at the time, widescreen was originally sold as "3-D without the glasses," because the screen has a slight curve that was thought to give an impression of depth. In practice this effect was negligible, but the wider format is certainly more natural to the eye.

Whenever a new innovation comes to the movies, there's a balance between its ability to draw a crowd and the cost to the movie theaters. This lesson was learned in the early days when the cost of converting to sound for the talkies pushed a fair number of theaters out of business. Ideally, these new widescreen movies should have been shot on film stock that was in the same ratio. But that would have meant new cameras, new projectors, and vast expense all round. Instead, the earliest widescreen was achieved by simply blanking off the top and bottom of the image on a standard piece of 35-millimeter film—throwing away part of the image to achieve a different shape.

This approach wasn't practical, as images got wider; the result was a disproportionate loss in vertical resolution as less and less of the frame's depth was used. In the end an anamorphic process was introduced. The idea is simple. A distorting lens, placed over the front of the camera, takes in a wider field of view than a normal lens, but squeezes the picture sideways so that it fits onto conventional film. The result is images that are squashed up sideways, with tall, thin sticklike people. When the movie is shown, a distorting lens is also placed in front of the projector. This lens stretches out the image sideways, meaning that widescreen pictures can be shown without a new projector. So where Muybridge had to elongate his images, because the zoetrope would then compress them, the widescreen movie industry had to compress onto the film so that theaters could elongate the final result.

The distortion Muybridge applied to make the zoetrope images look realistic was intentional and was clearly a photographic process, even though, as was inevitable at the time, there was some touching up:

> Much time and care were required in the preparation of the discs. . . . For many of the discs it was found advisable to fill up the outlines with opaque paint, as a more convenient and satisfactory method of obtaining greater brilliancy and stronger contrasts on the screen than was possible with chemical manipulation only. In the "retouching" great care was invariably taken to preserve the photographic outline intact.

Whether what was first projected was Muybridge's original photographic sequence or a painted version of it, the zoöpraxiscope was nothing less than a motion picture projector. It was very limited, with between 12 and 24 images to a disc, yet the fact remains that crude though they were, these were movies. In fact, in one respect, the Muybridge approach was better than the first traditional movies. Ernest Webster, who acted as operator for Muybridge on many of his lectures, commented that the zoöpraxiscope ran smoothly, not flickering like the early cinematograph and vitascope movies, which were notoriously jerky.

The relative smoothness reflects the different approach of the systems. Normal movie projectors have a rotating shutter, often shaped like a butterfly. When the opaque part of the shutter is over the light source, cutting off the illumination, the film is pulled along by a pawl that hooks into the sprocket holes along the side of the film or by a special sprocket that only moves when the shutter is in place. Applying this operation three times to each frame to take the shooting speed of 16 frames per second up close to the desired 50 images a second, plus mixed quality in the mechanisms of early projectors resulted in uneven projection. By comparison, though the zoöpraxiscope shutter had to be narrow to keep each image "frozen" in place, so though the image was noticeably weaker, the system had a smooth flow, devoid of the potential for jerkiness of the film projectors.

That the moving picture was something special, something different from a series of lantern slides was illustrated on the very first ever demonstration of the zoöpraxiscope at the house at Palo Alto in the autumn of 1879, a private showing for Leland Stanford and some friends. Muybridge set up the machine and announced that he was going to show them the horse Hawthorn in action. He got an appropriately delighted response from his small audience. Stanford himself, though, seemed more puzzled than impressed. "I think you must be mistaken in the name of the animal, Mr. Muybridge," he is reported as saying. "That is certainly not the gait of Hawthorn, but of Anderson."

When Muybridge checked with the stable staff it turned out that on this particular day they had substituted Anderson for Hawthorn but had not bothered to tell Muybridge. After all, for his purposes, one horse was much the same as another. Yet from the crude, shadowy images, Stanford had been able to tell he was watching a different horse. The moving picture did not just animate, it provided more information, more depth of content than a series of individual snaps. It wasn't just a different type of photograph, it was a new communication medium. Muybridge had prefigured both the movies and that even more pervasive channel of communication, television.

The effect that Stanford had unconsciously recognized was that moving pictures add extra dimensions to the information presented to the viewer, making it a richer medium than a still photograph. The most obvious enhancement is the addition of the time dimension to the two spatial dimensions on the print. It was the addition of time—making it possible for the motion and change sensors in the brain to come into play—that allowed Stanford to recognize that a different horse was in use. More subtly, moving pictures also added depth. Although any particular image is just as flat as a conventional photograph, because moving objects on the screen can pass behind and in front of other objects, the three dimensions of the environment are made clearer. Motion provided essential cues for Stanford's brain to recognize what was nearer and what was farther away.

The story above of Leland Stanford's observation is the version told by George Clark, Stanford's official biographer, though Muybridge himself told a slightly different story when lecturing at the Royal Society of Arts in London. According to the proceedings of the society, Muybridge said to Stanford that he supposed he would recognize his horse Florence Anderson.

> "Well," said Mr. Stanford, "You have got a galloping horse there, but it is not Florence Anderson."
>
> [Muybridge] replied that it was, that the trainer had sent it out to him and that was a picture of it. Mr Stanford said that he was quite satisfied that it was not Florence Anderson, and on the man being sent for, it turned out to be another horse.

Bizarrely, a third version of the events can be found in a huge article Muybridge wrote for the *San Francisco Examiner* of February 6, 1881, which provides yet another combination of horses:

> Mr Stanford looked at it.
>
> "That is Phryne Lewis," said Mr. Muybridge.
>
> "You are mistaken," said Mr. Stanford, "I know that gait too well. That is Florence Anderson."

As in each of the other versions, Stanford was able to recognize the horse from the way it moved on the screen—and in the end, that's the important part of the story.

It might seem surprising that Muybridge, who had been quick to establish his rights on the advances he made within the camera shutter, never took out a patent on the zoöpraxiscope, arguably his greatest contribution to history. It seems, however, that the Patent Office short-sightedly decided that Muybridge's device was not "sufficiently inno-vative" and rejected the application, entirely missing the huge difference between a moving picture projector and the zoetrope. Even though Muybridge's device threw a smooth image onto a big screen to entertain a mass audience, it was considered on a par with squinting through a set of rotating slots in a tin cylinder.

Undaunted, Muybridge was ready for a more open demonstration by the following spring. On May 4, 1880, the zoöpraxiscope had its first truly public outing at the San Francisco art association. Muybridge put his horses, animals, and athletes through their paces on a large screen. The press was effusive, spotting that this was the start of some-thing big:

> Mr. Muybridge has laid the foundation of a new method of entertaining the people, and we predict that his instantaneous, photographic, magic lantern zoetrope will make the rounds of the civilized world.

The *Alta California* didn't know the half of it when it came to the new methods of entertaining the people that would follow, and per-haps had a little to learn from Muybridge when it came to less clumsy product names—it's hard to imagine selling an "instantaneous, photo-graphic, magic lantern zoetrope" to anyone but a writer of comic songs. But they were not alone in their enthusiasm. The *San Francisco Call* commented:

> Nothing was wanted but the clatter of hooves upon the turf and an occa-sional breath of steam to make the spectator believe he had before him the flesh and blood steeds.

If this seems an overreaction to a dozen or two frames of simple movement on a screen, we should remember just how revolutionary the experience was. Up to then the best you could expect was a lantern slide with a moving part that was waggled back and forth by the operator. Here was what appeared to be reality brought back to life from a series of photographs.

Only one newspaper was negative at the time, suggesting that a local artist called Jules Tavernier gave Muybridge the idea of projecting his photographs in this way, rather than Muybridge thinking of it himself. There is no evidence that this was the case, and it seems to be a publicity attempt on Tavernier's behalf, rather than a serious claim.

For more than a year Muybridge consolidated his work. He published a collection of the key photographs from the Palo Alto experiments, including some extra shots taken especially for the book: a series in which a horse's skeleton was arranged in the various positions of the trotting sequence to show how the underlying bone structure worked. The resultant book, *The Attitudes of Animals in Motion: A Series of Photographs Illustrating the Consecutive Positions Assumed by Animals in Performing Various Movements,* was not mass produced. Although the title page, introduction, and contents were printed normally, the rest consisted of a series of 203 hand-printed photographs. Only five copies of this work were definitely made, a time-intensive process for Muybridge. Much of the rest of his time was spent lecturing and demonstrating the zoöpraxiscope, though he did find time to patent the variant on his shutter release that allowed a series of exposures to be made on a timed basis, as had proved necessary to cope with capturing the antics of the less predictable animals he had photographed.

Come late summer of 1881, though, Muybridge was ready to face a wider audience. He set off for Europe, taking with him the zoöpraxiscope, many photographs, and a small number of copies of *The Attitudes of Animals in Motion.* The cameras from Palo Alto, which Stanford appears to have given to Muybridge as part of his payment for the work there, he stored in New York. He stopped briefly in London (and Kingston), but his first main port of call was Paris.

⚬⚬⚬

Muybridge had kept up a correspondence for several years with Professor Étienne Jules Marey, and it was he who acted as Muybridge's minder while in France. The first demonstration of the zoöpraxiscope was in Marey's home, where it drew a spectacular response from the audience. Marey acted as an interpreter, giving a commentary for Muybridge, whose French was almost nonexistent.

The Paris newspaper *Le Globe* described the event in the Boulevard Delessert, where "some foreign and French savants" gathered to see "the curious experiments of Mr. Muybridge, an American, in photographing the movements of live creatures." The article concluded with a tongue-in-cheek recommendation that shows that frustration with the slowness of public transport is nothing new:

> Finally, let us address an inquiry to Messieurs Marey and Muybridge. Would it not be practicable by the zoetrope process to give a little additional speed to the Paris carriage horses? We should not require elegance of movement, but hurried journalists would be under lasting obligations to the inventor of some such contrivance.

Perhaps the most significant date in Muybridge's highly successful visit to Paris was October 26, 1881. A gathering of over 200 of the great and good of the Paris art world, including Millet, Bonnat, and the playwright Alexandre Dumas fils (son of the great Dumas and a popular author in his own right), were entertained in the mansion of the popular painter Jean Louis Meissonier at 131 Boulevard Malesherbes. The then-famous artist seemed to look on Muybridge as a protégé. He translated for him and emphasized to the gathering what an amazing resource for artists Muybridge had provided. This was no chance enthusiasm, but rather the outcome of some excellent political maneuvering on the part of Leland Stanford.

Stanford had been in Paris in 1879, in part to get portraits of his wife and himself painted. He arranged with no problem to have Bonnat paint Mrs. Stanford, but came up against a problem with his own choice of artist—Meissonier. Stanford was used to getting his own way. It would probably not be exaggerating too much to say that he had an attitude that said that anything he wanted could be bought—and he had the cash to buy it. However, Meissonier took his art seriously and was successful enough to be able to turn down commissions. When Stanford asked to have his portrait painted, Meissonier responded with a flat Gallic, "Non."

Stanford now proceeded to make matters worse for himself by brandishing his checkbook and asking Meissonier to name his price.

He had critically miscalculated the nature of the man. For Meissonier this was little more than an invitation to prostitution. The only outcome was to increase the stubbornness with which Meissonier refused.

Stanford, however, was not a one-strategy player. Seeing that Meissonier could not be bought in the conventional way, he decided to play on a different weakness. Looking around the studio, he admired Meissonier's paintings of horses, a subject of which Meissonier considered he had absolute command. He had long studied the motion of horses to get his paintings perfect, repeatedly sketching animals in motion as he rode alongside them in a carriage.

Stanford asked if Meissonier, with his remarkable knowledge of horse movement, could sketch a horse in a trotting motion, making the legs look realistic. This request seemed perfectly acceptable to Meissonier. He threw off a simple but impressive quick sketch. Here was the starting point Stanford needed. He acknowledged Meissonier's skill, but he wanted just a little more. He asked the artist to repeat the sketch, but to do it when the horse had moved another 12 inches along the ground, so its legs would all be subtly differently positioned.

Initially, the task didn't seem too much of a challenge to Meissonier either. He started a second sketch, but very rapidly his confidence dropped away. He rubbed the pencil lines out and started again, and again, and again. Finally, he had to admit he was not capable of doing what Stanford had asked. Yet.

Now was the time for Stanford to play his trump card. He pulled from his bag a collection of the Muybridge photographs. Meissonier was dumbfounded. He had seen a few photographs at Marey's laboratory but had not been hugely impressed. This, though, was something different. With hardly a second thought he now promised Stanford his portrait, presumably in exchange for access to the photographs, though the artist would also later say that he spent a sleepless night going over and over what he had seen, and swore to himself that he would never paint another horse. Meissonier's own account of his meeting with Stanford does not include the challenge, but his reaction to Stanford, brought to the studio by "some American dealer," is clear:

> Some American dealer, I have forgotten which, brought a certain Mr.
> Leland Stanford, and his wife, to my studio. He asked me to paint his por-

trait. My first impulse, of course, was to refuse, but he began to talk about the photographs of horses in motion, and said they were his. He had even spent $100,000 on the work, so a friend who was with him said, and the proofs which had reached Europe were a mere nothing.

Meissonier did paint Stanford's portrait; in it Stanford looks as if he is shifting uncomfortably in his armchair. Pinned under his left elbow is an open copy of Muybridge's book. As part of his fee Meissonier demanded that Stanford fund a trip for Muybridge to come to Europe. Whether Muybridge knew that Stanford's generosity was designed to elicit a portrait from the famous painter is not clear. But Muybridge was not one to look a gift horse in the mouth.

The Stanfords made a return visit to Paris while Muybridge was there, but it turned out to be a short and unsatisfactory trip. Stanford was ill and, as Muybridge put it in a letter to Frank Shay, superinten-

Meissonier's portrait of Stanford.

dent of Stanford's Palo Alto stock farm, on November 28th, his "residence in Paris has been entirely devoid of pleasure, both to himself and Mrs. Stanford." The letter portrays the enthusiastic reception that Muybridge was receiving and goes on to describe his plans for future developments (the underlining in this and other letters quoted was Muybridge's own):

> I shall shortly visit England for the purpose of inducing some wealthy gentlemen (to whom I have letters of introduction) to provide the necessary funds for pursuing and indeed <u>completing</u> the investigation of animal motion; and in framing an estimate or the probable cost, can have no better basis than the cost of the work already accomplished.
>
> Will you therefore at your <u>very earliest convenience</u> favor me with the total amount of money paid to me, or on my account.

With this, the real reason for the letter is revealed. He needs some information and he needs it quickly. Muybridge goes on to request such detail as "Cash paid to Muybridge for personal use NOT <u>including</u> the $2000 the Gov. gave me." He concludes as a postscript, "Please don't delay sending statement."

Less than a month later Muybridge wrote to Shay again. He still wanted those figures, but it seemed that a concrete proposal was emerging more quickly than he had expected, and not in England, but in France. He now hoped to intercept the figures, which he had asked to be sent on ahead of him:

> I suggested addressing your reply to London, but since then some important events have transpired which will render an extended stay in Paris necessary; and at the same time relieve me of the anxiety under which as you well know I have for a long time been existing.

An impressive grouping of the scientist Marey, the artist Meissonier, Muybridge, and an unnamed financier had put together a plan to undertake the next stage of the experiments, using the latest photographic technology, with the aim of producing a book that would explore motion in art. Muybridge was eager to let Stanford know of his plans, probably to make sure that he would not throw up any difficulties, though he would also not have been averse to Stanford making a financial contribution. This he had apparently already brought up with Stanford, as he writes:

> Both [Meissonier] and I consider it appropriate to invite the Governor to join us if he is so disposed, which we have done by letters, we shall be pleased to welcome him if he is inclined to come in, if he declines we will avail ourselves of M. Meisonnier's friend.

According to Muybridge, the finance wasn't coming from Meissonier because

> notwithstanding the large prices obtained from his pictures, unfortunately M. Meissonier is far from rich; but his influence with wealthy people is immense; and one of his friends has expressed a desire to associate himself with M. Meissonier, Professor Marey and myself in the instituting of a new series of investigations which I intend shall throw all those executed at Palo Alto altogether into the shade. I have been experimenting a great deal and have no doubt of its successful accomplishment.

In fact, M. Meissonier was pretty well off, but perhaps he did not want to put his money into the venture. No matter; they had a financier. Muybridge did not show himself to be much of a diplomat in this letter. A man with Stanford's powerful self-esteem would hardly be thrilled to learn that the Palo Alto experiments, so strongly associated with his name, were to be thrown "altogether into the shade"—especially by Europeans. Almost certainly unintentionally, Muybridge was firmly severing his connection with Stanford with this letter.

Unfortunately, though, it seems that the mystery financier pulled out. Muybridge had clearly expected to stay in Paris—he finished his letter to Shay with a strangely inelegant invitation to come and visit:

> I hope your mine has turned out enough rich paying ore to satisfy the reasonable requirements of any moderate man, and that its results will enable you to retire from speculation, and seek employment for a time in Europe; and if in the course of your travels you should next summer find yourself in Paris; make me a visit to my Electro-Photo studio in the Bois de Boulogne and I will give you a welcome.

But unspecified delays seem to have made it necessary for Muybridge to travel to London after all. The new project was not forgotten—he had hopes of getting funding from one of the London institutions—but it was for the moment put on ice.

By early 1882, favorable reports of Muybridge's reception in Paris had reached England. In March of that year he was to take another country by storm. But a very different kind of storm cloud was looming over his relationship with Leland Stanford.

⟨⊚⟩

The 13th of March 1882 found Eadweard Muybridge lecturing at the Royal Institution. This was a significant honor. Of the two great British scientific establishments, the Royal Society may have more social prestige, but the Royal Institution was the home of practical science. Leading lights of the Royal Institution had included Sir Humphrey Davy and the incomparable Michael Faraday. To be invited to lecture there was a mark of acceptance in the hallowed halls of technology.

The event was a sell-out success. A repeat performance had to be arranged for later in the day so that those who could not squeeze into the auditorium could see the wonders on display. And the audience was one that Muybridge could hardly have dreamed of mere months before. Not only was the lecture hall stuffed with notables from both the Royal Institution and the Royal Society, not only were there dignitaries such as the poet laureate, but his lecture was also graced with a royal presence. The prince of Wales (who would become King Edward VII), the princess of Wales, their daughters, the princesses Alexandra, Victoria, Louisa, and Maud, plus the duke of Edinburgh, were all in the audience.

According to an excited reporter for the *Photographic News,* "Mr Muybridge might well be proud of the reception accorded him, for it would have been difficult to add to the éclat of such a first appearance, and throughout his lecture he was welcomed by a warmth that was as hearty as it was spontaneous."

It is worth following the article in some detail as it gives a clear description of how Muybridge's lectures—performances he would eventually go through hundreds of times—were structured. We are told that he left his "wonderful pictures" to speak for themselves, instead of lecturing at length, showing the photographs on the screen and "modestly [explaining] them in clear but plain language." This was no tedious speech with the occasional slide; the whole thing was built very much around the visuals, the thing the audience, after all, had come to see.

Muybridge began by describing the track and equipment, with photographs of the site to explain how the pictures were taken. Then he moved on to the subjects themselves:

Mr Muybridge, by way of comparison, first threw on the screen a series of artists' sketches of the horse in motion, some of them old-world designs of the Egyptians and Greeks, some very modern, including the principal animal from Rosa Bonheur's well-known "Horse Fair." In no single instance had he been able to discover a correct drawing of a horse in motion, and, to prove his statement, he then threw on the screen several series of pictures representing the different positions taken up by a horse as he walks, trots, ambles, canters or gallops. One thing was very plain from Mr Muybridge's pictures, namely, that when a horse has two of his feet suspended between two supporting feet, the suspended feet are invariably lateral; that is to say both suspended feet are on the same side of the animal. This, no painter—ancient or modern—had ever discovered.

Muybridge went on to show that the horse's different gaits involved entirely different movements of its legs. By now, nonspecialist members of the audience might have been slightly losing interest, and it's possible to imagine the younger royals getting rather fidgety, but Muybridge was too good a showman to lose them for long. He then introduced the zoöpraxiscope.

After Mr Muybridge had shown his audience the quaint and (apparently) impossible positions that the horse assumes in his different gaits, he then most ingeniously combined the pictures on the screen, showing them one after another so rapidly, that the audience had before them the galloping horse, the trotting horse, etc. Nay, Mr Muybridge, by means of his zoepracticoscope [sic] showed the horse taking a hurdle—how it lifted itself for the spring, and how it lightly dropped upon its feet again. This pleasing display was the essence of life and reality. A new world of sights and wonders was, indeed, opened by photography, which was no less astounding because it was truth itself.

Writing for a photographic magazine, the author made some comment on the fact that all the existing photographs were made on old wet plate technology, and with the latest advances in the science, even more could be achieved:

Mr Muybridge modestly calls his series of animals in motion—they include horse, dog, deer, bull, pig etc.—simply preliminary results. They contain little or no half-tone and are only proof of what may be done. What he desires now to secure, if he only receives sufficient encouragement, is a series of photographic "pictures," and these, with the experience he has now acquired, and with the gelatine process to help him, should be well within his reach. We only trust this encouragement will be forthcoming, and that Mr. Muybridge will be tempted to carry on the difficult work he has commenced with such genuine success.

In other words, the pictures were largely black and white with no shades of gray—the next step was to get full detailed motion photographs, which would be helped by the new gelatin-based dry plate technology that was sweeping through the photographic business, making it much easier to take consistent, rapid photographs. As for the encouragement that the writer hoped would be forthcoming, it seems to have been a code word for cash.

Perhaps the most fascinating aspect of the event was the prince of Wales's intervention. He put his finger, albeit unwittingly, on two of the key applications of motion picture technology—to entertain and to show sports to those who weren't present at the time. The prince asked Muybridge if they could "see your boxing pictures." These were some of the athletes from the Olympic Club that Muybridge had photographed at Palo Alto when not working with animals.

Muybridge missed the commercial opportunity, responding, "I don't know that these pictures teach us anything very useful, but they are generally found amusing"—something of an understatement considering the future of sports on film and TV. When the boxers were shown, it was reported as being "to the infinite delight of the audience in general and the prince of Wales in particular."

A more dramatic version of events was found in the more sensationalist *Illustrated London News*, published five days after his lecture. In it we hear that "by the aid of an astonishing apparatus called a 'Zoöpraxiscope,' which the lecturer described as an improvement on the old 'zoetrope,' but which may briefly be defined as a Magic Lantern Run Mad (with method in the madness), the ugly animals suddenly became mobile and beautiful, and walked, cantered, ambled, galloped and leaped over hurdles in the field of vision in a perfectly natural manner."

The reporter goes on to say, "I am afraid that, had Muybridge exhibited his 'Zoöpraxiscope' three hundred years ago, he would have been burned as a wizard."

The day following the sell-out success at the Royal Institution, Muybridge, as if to stress the way that his work spanned the usually opposing factions of science and art, spoke at the country's leading artistic body, the Royal Academy. One of the cuttings that Muybridge saved in his bulging scrapbooks covers the event, and is particularly interesting in that it was written by the London correspondent of one of the San Francisco newspapers, the *Morning Call.*

He starts by describing Muybridge's unparalleled reception in London, commenting that "in all my long experience of London life I cannot recall a single instance where such warm tributes of admiration for merit have been unstintingly given by the greatest of the land."

He then goes on to describe Muybridge's reception at the Royal Academy, seemingly encouraged by the event into a small fit of rococo descriptiveness:

> To me, I confess, it was a sight of pride to behold a countryman of mine installed as an instructor within the sacred walls of the great temple of fine arts, there to give the fruits of his knowledge and the sum and substance of his investigations to such listeners as Sir Frederick Leighton, Alma-Tadema, Marks, Orchardson, Vicat Cole, and in fact the whole body of the R.A.s.

This was London's artistic establishment out in force. The journalist sat next to the distinguished Frederick Leighton (bizarrely now remembered as Lord Leighton, although he only received the baronetcy on the day of his death), whose classically influenced work was photorealistic in its detail. The journalist noted how Leighton was fascinated by the surprising positions of the horse's legs, "[imitating] it with his hand and arm in explanation to the various artists gathered near him."

The *Morning Call* reporter goes on:

> Certain it is that if any artist, before the advent of Muybridge's photographs, had the temerity to paint horses in the positions which the man from Palo Alto has shown to be those which these animals assume when in motion, that same artist would have been reproached for curious drawing. Now I should not be in the least surprised if next year's batch of paintings at the Salon and Royal Academy were to show the influence of Mr. Muybridge's photographs in the most pronounced manner. Meissonier told Muybridge that he had given him entirely new lights on the position of horses.

After commenting on the enthusiastic response of Sir Lawrence Alma-Tadema (a Dutch-born painter, also with a realist, classical-inspired style, but some of whose work has since been considered little more than high-class pornography) and Leighton, the reporter goes on to emphasize Muybridge's new fame, and to compare him very favorably with an unnamed Californian author. There are a number of possible targets the reporter could have had in mind. Most likely it was Mark Twain, as, according to a contemporary article, despite the fact that Samuel Clemens was born in Florida it was "the fashion to call Mark Twain a Californian writer."

"Thus," remarks the *Morning Call* reporter, "Mr. Muybridge is launched upon the sea of London celebrity. Let us hope he will not be spoiled by it, as a certain California literary man has been. I do not fancy Muybridge is one of the kind who gets spoiled by flattery. There seems too much hard sense and practical thought there for anything of this sort."

The reporter then does some thinking about the potential for commercial use of Muybridge's work. He reckoned that if Muybridge were to hire a hall and give exhibitions twice a day at a shilling entrance fee, he would "clear enough money during the coming summer to greatly assist him in the pursuance of the researches in the field where he has already made such curious and unexpected discoveries."

There is no doubt he was right, though it would be a good few years before Muybridge would earn money directly from paid shows. The writer concluded with a burst of pride for his California colleague and his "true American courage." Bearing in mind that Muybridge was a son of Kingston-upon-Thames in Surrey, and never took American citizenship, this was more a case of enthusiasm than accuracy.

Although Muybridge did not hire a hall, he did continue to give a series of lectures in distinguished locations from South Kensington to Eton College through to June 1882. He made it clear that he was looking for support in furthering his experiments, but as yet, though the response to his lectures was uniformly one of delight and enthusiasm, he was not beset by potential sponsors. It is surprising that there weren't more millionaires beating down his door—this was popular stuff.

The two great bodies, the Royal Institution and the Royal Academy, had already demonstrated their enthusiasm for Muybridge and his work; it only remained for the last and perhaps most renowned of the triumvirate, the Royal Society, to celebrate his work. Already active in Newton's day, the Royal Society was probably still in Muybridge's time the foremost discussion forum for scientific work in the world. The president of the society, the mathematician William Spottiswoode, who had been at the Royal Institution lecture, invited Muybridge—a man who had no academic honors—to submit a monograph on animal locomotion.

What happened next can be told in Muybridge's own words from a draft of a letter to Leland Stanford, dated May 2, 1892, found among Muybridge's papers. It is not known whether the letter was ever sent, but the mere fact that he wrote it 10 years later showed what a profound impact the events at the Royal Society had on Muybridge. In the letter he tells Stanford:

> The monograph was examined, accepted and a day appointed for its presentation to the Fellows, and for its being placed in the records of the Society.
>
> I have in my possession a proof sheet of my monograph printed by the Society, (as is its custom) before being place [*sic*] on the record of its "Proceedings."
>
> About three days before the time appointed for the reception of my monograph by the Fellows, I received a note requesting my presence at the Rooms of the Society.
>
> Upon my arrival I was conducted to the Council Chamber and was asked by the President in the presence of the assembled Council, if I knew anything about a book then on the table having on its title page, the following, there being no reference thereon to Muybridge:

In his handwritten letter Muybridge had drawn a heavy box around the following words, reminiscent of the black borders that were placed around letters of condolence:

THE HORSE IN MOTION

by

J. D. B. STILLMAN MD

Published under the auspices of

LELAND STANFORD

Stunned, Muybridge leafed through the book in front of the Council of the Royal Society. This was not his elegant record of the Palo Alto experiments; it was an entirely different book, printed with engravings, rather than containing original photographs. It contained a long description of the work at Palo Alto and around 90 illustrations based on Muybridge's photographs—but he was not given credit. Muybridge continues:

> I was asked whether this book contained the results of the photographic investigation of which I had <u>professed</u> to be the author. That admitted, I was invited to explain to the Council how it was that my name did not appear on the Title Page, in accordance with my profession.
>
> No explanation of mine could avail in the face of the evidence on the title page, and in the book before the Council, I had no proof to support my assertions. My monograph was refused a place on the records of the Royal Society until I could prove to the satisfaction of the Council my claim to be considered the original author, and until this day it remains unrecorded from lack of evidence which would be acceptable to the Council, which evidence is at your command.
>
> The doors of the Royal Society were thus closed against me, and in consequence of this action, the invitations which had been extended to me were immediately cancelled, and my promising career in London was thus brought to a disastrous close.
>
> My available funds being exhausted I was compelled to sell the four original photographic copies of "The Horse in Motion" which I had printed at your request, and for your purposes, and with the proceeds of their sale I returned to America.

These were the beautiful, handmade copies of *The Attitudes of Animals in Motion* that Muybridge had taken with him to Europe.

During a recent cleanout of the archives at the Royal Society, a fascinating postscript to this shocking occurrence was discovered, and now, for the first time, it is possible to understand exactly what happened back in 1882. Although Muybridge's paper never made it into the transactions of the Royal Society, they did keep on record the original notes on his paper that triggered Muybridge's humiliation in front of the council.

These hurried, scribbled comments come from an illustrious, if now controversial, hand—Sir Francis Galton—no ordinary scientist. Galton began his career as an explorer, penetrating parts of Africa at the time unknown in Europe, and was also well known as a meteorologist, but he was to undertake a significant change of direction, influenced by Darwin's *Origin of Species*. Galton became fascinated with the early attempts to understand genetics and heredity, and it was from this fascination that his fame—and notoriety—would grow.

By 1882, when he reviewed Muybridge's paper, the 60-year-old Galton was a leading light in the field of heredity, and that was how he would have been seen at the time. By then, however, he had already started on what some would interpret as a darker path. In his 1869 book *Hereditary Genius*, Galton had introduced a new term "eugenic." The concept of eugenics was one he was to increasingly focus on and popularize. This is the apparently innocent seeming aim of improving the human race by selective breeding—the sort of thing that race horse owners like Stanford did as a matter of course.

Galton's present-day apologists argue that there was no harm in his work. On the very full website dedicated to Galton (www.galton.org) the editor Gavan Tredoux comments, "Contrary to modern superstition, [eugenics] is an entirely respectable cause if pursued sensibly, practically and ethically." Unfortunately, as regimes like Nazi Germany's have proved, pursuing eugenics ethically is easier said than done. The very idea of improving the human race by a selective breeding program, even without the horrors that have accompanied historical attempts is repugnant to many.

This then was the man who reviewed Muybridge's paper. It might seem that he would inevitably have been biased in favor of Stanford—a man obsessed with the scientific improvement of the equine race—yet that would be an unfair assumption; there is no evidence that Galton's science was anything other than scrupulous. And, while it makes for a neat conspiracy theory, an early part of Galton's report makes it clear that he did not know Stanford:

> Mr. Muybridge's memoir is founded on a numerous series of "instantaneous" photographs made by him in California (at the cost of [illegible amount]) at the station of a wealthy cattle breeder, Mr. Leland Standford [*sic*].

No one who knew Stanford would refer to him as a cattle breeder. However, while there is no suggestion that Galton was in bed with

Stanford, it is clear that his only interest in the work is biological. This was why he was chosen to review the paper in the first place, and this was entirely Muybridge's fault, as his paper "Attitudes of Animals in Motion" concentrates more on what can be learned about the gait of horses than the indubitable breakthroughs in photographic capability. Galton goes on to put the paper into context:

> There is a dispute as to the scientific proprietorship of these photographs. Mr. Stanford considers Mr. Muybridge as being in his employment, Mr. Muybridge claiming a separate right to his own work. However that may be, Mr. Stanford has already published the photographs in an elaborate volume, together with a description of the conclusions to be drawn from them, which latter was written by another employee of his, Dr. Stillman.

Galton may or may not have intended it, but it is clear that he had already made up his mind about Muybridge's status. Note that he says the book was written by another employee—he had decided exactly what Muybridge was: an employee on the make. Galton's was a wealthy middle-class background, and his natural inclination would be to see things from the point of view of the employer. He was also technically inaccurate in his description of the Stillman book, which contained not Muybridge's photographs but less effective drawings based on his photographs. Next came the killer paragraph. Galton compares the content of Muybridge's paper and the Stillman book:

> I do not find that the conclusions drawn by Mr. Muybridge in his memoir differ in any notable manner from those of Dr. Stillman; they are certainly much less perspicacious and well expressed, and they are scarcely if at all intelligible without reference to the photographs. If then Mr. Muybridge's paper were published, it would have to be illustrated with reproductions of the photographs to which he has no sure legal claim, and after this was done the result would be a memoir, much inferior in importance to an already published book, without perpetrating any novel feature of its own, that I can see, to recommend it.

Galton seems to have reached a strange conclusion. Stillman's text was stilted and uninspired. Muybridge's paper was much more readable to the modern eye—it may be that the very stiltedness of Stillman's prose was more acceptable to the Victorian taste. Galton does make a fair point that Muybridge's paper really needs the photographs to support it, but then falls down by stating "to which he has no legal claim"—an assertion that seems to be made without any attempt to check the facts, as Muybridge had registered copyright of the photographs. The conclusion Galton reached should hardly be a surprise by now:

> Under these circumstances I am unable to advise the council of the Royal Society to order Mr. Muybridge's paper to be presented either in their Transactions or in their Proceedings.

> Francis Galton

> May 19 / 82

As far as Muybridge knew he was damned by the absence of his name on the title page. He never saw this assessment, which until now has been locked away in the vaults of the Royal Society. He was not given the chance to prove his ownership of rights to the photographs nor to argue his case because he was unaware of the detail of the report.

Even accepting Galton's doubtful argument that Stillman's text was better than Muybridge's, the logic is strange. Imagine that a science journalist had written a book about some exciting new research going on in a university, and managed to get the book published before the scientist's paper reached a journal. It's hard to imagine any editor refusing the paper because "there's already a book about it." The fact is, Stillman had not even been present when the research was undertaken. Muybridge's paper had every right to be accepted.

It is interesting that Galton, after so roundly dismissing Muybridge's contribution, only one month later in June 1882 produced a series of composite photographs from Muybridge's originals in a montage titled "Explanation by Composites of Muybridge's Photographs of the Conventional Representation of a Galloping Horse," in which he explains how the eye is fooled by the high-speed motion of the horse's legs to produce the legs-splayed look so common in horse paintings. It seems that he was prepared to take a quite different attitude in public to Muybridge's work to that in his secret review of the paper submitted to the Royal Society.

Muybridge's horror at his treatment was not just a matter of the irritation of not getting a paper published. The Royal Society was no mere journal publisher; it was at the heart of the moneyed scientific establishment. Any hope Muybridge had of inducing some wealthy benefactor in England to fund the completion of his work had been dashed. What gentleman would pay money to a man that the Royal Society appeared to be branding a fraud? Muybridge's dream project of cooperating with Marey and Meissonier was shot to pieces.

But how did this book turn up to shatter his hopes? How, when his fame and fortune seemed assured, could things have gone so horribly wrong?

NINE

CONDEMNED ONCE MORE

I have for a long time entertained the opinion that the accepted theory of the relative position of the feet of horses in rapid motion was erroneous. I also believed that the camera could be utilized to demonstrate that fact, and by instantaneous pictures show the actual position of the limbs at each position of the stride. Under this conviction I employed Mr. Muybridge, a very skilful photographer, to institute a series of experiments to that end. Beginning with one, the number of cameras was afterwards increased to twenty-four, by which means many views were taken of the progressive movements of the horse. . . .

When these experiments were made it was not contemplated to publish the results; but the facts revealed seemed so important that I determined to have a careful analysis made of them. For this purpose it was necessary to review the whole subject of the locomotive machinery of the horse. I employed Dr. J.D.G. Stillman, whom I believed to be capable of the undertaking.

From Leland Stanford's preface to Stillman's
The Horse in Motion

The existence of the book that caused such disgrace and shame for Muybridge was not a surprise to him. He was well aware that it was being written, but had taken little interest in it, fairly naturally assuming that he would be given the appropriate credit for his work.

Dr. J. D. B. Stillman (a medical doctor rather than a veterinarian) was a friend of Leland Stanford's, who at Stanford's request had made a physiological study of horse movement to fit alongside Muybridge's photographic work. Back in 1881, before Muybridge left for Europe, Stillman had let Muybridge know that he intended to produce a book based on the work at Palo Alto, with which Muybridge had seemed quite comfortable.

Muybridge had a busy time in Europe—the book must have seemed a distant, humdrum venture by comparison with the excited audiences at his lectures. He did write to Stanford from Paris when he was introduced to the superior dry plate photographic equipment, suggesting that they hold up the book's publication so that some dry plate images could be included, but this seems to have been ignored.

Dry plate technology would quickly transform photography. As we have seen, the wet collodion process was difficult, time consuming, dangerous, and particularly impractical away from the studio. Almost as soon as it was devised, attempts began to find an alternative that would work with dry materials, meaning that a plate could be carried already prepared and used when and where required. In 1871, an English physician who dabbled in photography, Dr. Richard Leach Maddox, found the answer when he realized that the ether vapor from his collodion plates was damaging his health—instead of collodion, he used gelatin.

This protein, extracted from animal matter and used in the catering business in products like Jell-O, proved an ideal substitute for collodion as the material in which to suspend the silver crystals. It meant that a plate could be produced commercially and bought and used as and when required. Initially, the dry plates were less sensitive than collodion, but soon Charles Bennett, the first to commercially exploit the process, and others found that it was possible to increase the sensitivity of dry plates by heating the sensitized gelatin over a long period.

This produced a plate that could produce an image in a tenth of the time required by collodion, making it an ideal medium for high-speed photography. The potential to make this speed increase arose from the difference in the way the silver iodide and bromide crystals were suspended in the medium. Wet collodion started with iodides and bromides dissolved in the collodion. When the mix was sensitized with silver nitrate, the outcome was very fine grains of silver salts distributed throughout the material. Potentially this allowed for a lot of detail, but the fine grains were slow to respond to light.

In gelatin the silver crystals were larger, and the slow heating process devised by Bennett resulted in those crystals growing (a technique known as ripening). With bigger grains, it took significantly fewer photons of light to produce a blackened crystal—the plates were more sensitive, if less detailed. Many early photographers not equipped with Muybridge's high-speed shutters found the dry plate technology too fast; they were used to making an exposure by taking a hat off the camera lens, counting up to three, and replacing it. Dry plates would drive the development of more effective mechanical shutters.

The gelatin medium would also prove the catalyst for the move away from using plates altogether. George Eastman began in the photographic business in 1878, when he devised a machine for evenly coating a glass plate with the gelatin matrix, but it was discovered that the gelatin layer could also be applied to a continuous strip of celluloid. Celluloid is a material related to collodion, invented in the 1840s and first commercially employed by the American inventors, the Hyatt brothers, as a cheaper alternative to ivory for making billiard balls. Celluloid is a mix of collodion and camphor subjected to high pressure that produces a plasticlike material. A thin sheet of celluloid provided an excellent base for photographs, though it did have a tendency to catch fire or explode unless properly stored. The development of the flexible celluloid opened up the possibility of roll film (and movies that lasted longer than a few seconds) that would transform photography. As Muybridge explored the wonder of dry plates, though, this technology was still in the future.

∞∞

The next time we know that Muybridge gave any thought to the book on his experiments was in a letter to Stillman written on March 7, 1882, less than a week before his triumphant Royal Institution lecture. He wrote:

> You are I suppose still writing away: You perhaps recall what I originally told you about the time it would take; and, if you succeed in getting the work in the market before 1883 I shall consider you very fortunate. Who have you arranged with to publish it?

Not having heard anything for such a long time, Muybridge seems to have assumed that Stillman was merely writing a descriptive book without illustrations (though still based on Muybridge's work) as he comments:

> It was contemplated at one time to make use of my photographs for illustration, but having heard nothing further in relation to it, and from conversation with Mr. Stanford last Fall I suppose the idea of making arrangements for pictorial illustration has been abandoned.

Muybridge hoped that rather than rush into publication with what was available from the Palo Alto trials, Stillman would wait to get the fuller results from Muybridge's planned (but never actualized) collaboration with Marey and Meissonier:

> I anticipate no difficulty in pursuing the investigation on a large and more comprehensive scale than has been done and to an exhaustive conclusion. . . . I suggested awaiting the publication of any theories based on my work, until this was done as I was anxious all criticism should await the completion of the new experiments. I am promised every facility for work in Paris, but whether I shall commence there or in England I have not yet fully determined.

What Muybridge didn't know was that not only had Stillman finished writing, but the book was already in print in the form that would cause him so much embarrassment before the Council of the Royal Society. Even if *The Horse in Motion* had been further away from publication, from later letters Stillman wrote it seems likely he would have ignored Muybridge's offer. Stillman thought of Muybridge as nothing more than a hired hand with a camera, brought in by Stanford to do a job.

It may have been that Muybridge underestimated Stillman's writing speed, but equally it is possible that he suspected that the technical problems of reproducing photographs well in a printed work were holding up progress. Admittedly he had remarked that he assumed they weren't using his photographs, but this might merely have been an attempt to fish for a response from Stillman. The photographs were, after all, the key to the research.

What Muybridge didn't know, though, was that Stillman had, in effect, cheated. Apart from a handful of the photographs, the illustrations in the book were actually taken from drawings based on the Muybridge photographs. These images would be much easier, quicker, and cheaper to reproduce with the technology available at the time.

It wasn't until the 1890s that photographs would regularly be re-produced as a normal part of the printed page; in the 1880s the most common approach to getting a true photograph into a book was still to paste pictures individually onto the pages, a costly and time-consuming exercise. In most cases images in books would be based on some form of engraving. Book illustrations were much more art than craft for most of the nineteenth century. Three technologies were in use, all variants on the same theme.

The simplest approach was the woodcut. Here an artist would carve a negative image of the desired picture into a block of wood, which would then be used to stamp out the result, like making a potato print. Usually the wood blocks were incorporated with movable metal type to provide the overall image of a page.

Some artists were so proficient at this back-to-front thinking that they could carve directly onto the wood, but most worked from a nor-mal positive sketch as a master. Often there would be three individuals involved: one to make the original drawing, a second to copy it in re-verse onto the surface of the wood, and a third to carve out the block. The specific type of wood used to make the block would depend on the image required. This contemporary explanation comes from the children's magazine *The Little Corporal* in 1866:

> "What kind of wood do they use for engraving?" asks a country boy, who knows the name of every family of tree in the forest. Well, if they want to make an engraving for a hand-bill, such as you see for a circus or menag-erie, they would, perhaps, use pine, or, at best, mahogany. Such engravings are very coarse, and do not need fine lines. If they want to make a better picture, for a newspaper or magazine, they generally use box-wood, some-times pear, and, occasionally, apple and beech woods.

A second, more sophisticated, approach was the etching. A steel plate was coated with a mixture of wax and tar, which once set was blackened with candle soot. The original drawing was traced using a pencil, and the image transferred from the tracing paper to the pre-pared surface of the steel plate by passing it through a press. The etcher would then have a series of graphite markings on the surface to work from, appearing as a filigree of silver lines on black. Steel needles were used to take away the resisting mixture along the lines; the whole plate was then dipped in nitric acid, which ate away the exposed steel, leaving depressed areas on the plate to provide the master for printing.

The third technology, in origin closely related to etching though soon to diverge, was lithography. As the name's Greek origin (stone writing) suggests, here the medium was originally stone. To begin with, the idea was similar to etching. A stone block with a smooth surface would be partly covered with a resisting material, and an acid would then be used to eat away the exposed areas, leaving raised portions to print. Before long a more subtle and much more flexible approach was taken. The image required was drawn on the stone surface in a greasy material, and then the surface was made wet. The water naturally accumulated on the bare stone and was repelled by the greasy drawing. The surface was then rolled with a greasy ink. The pigment tends to accumulate on the greasy parts but was repelled by the water on the bare stone, so it is the greasy drawing that became inked and would print an image.

Stone tablets aren't a very practical medium. From about 1820, it was found that zinc and later aluminum and copper plates would work as well as the stone surface, but were much more practical. Crucially, such plates could be bent into cylindrical form, as the high-speed print technology developed during the nineteenth century moved from printing with flat single-press printing machines to a rotating drum that could pass through a continuous stream of paper. Variants of the lithographic process are still in wide use today.

When photographs were first reproduced in print, other than by pasting individual photographs into a book, it was by simply copying or tracing the photograph to produce an etching or lithographic image—this was the approach taken with most of the images in *The Horse in Motion* book—but this clearly was not satisfactory to illustrate the outcome of a photographic experiment. Unlike a drawing, a photograph did not consist of a series of lines, so could not be imposed directly as sections of greasy material. By the time Muybridge's next collection would go into print as *Animal Locomotion*, there was a delicate process for incorporating photographs into print called collotype.

The collotype was a halfway house between a true photograph and a normal printer's block. A photographic negative is rested on a sheet of glass coated with gelatin that has been sensitized with potassium or ammonium dichromate. When light is shone through the negative, the gelatin receiving the most light dries out. This gelatin-on-glass plate is then inked, with the ink repelled by the wetter sections and clinging to the drier gelatin, to provide a printing plate.

It was not until the 1880s that the American inventor Frederick Ives found a practical way to get a photographic image onto a normal lithographic plate. The process, known as halftone, converts the photograph into a series of dots that can be printed using a standard lithographic press. To make these dots, two evenly exposed, hence blackened, glass photographic plates were used. One plate had lines scribed on it horizontally, the other vertically. These two plates were stuck together so the lines formed a grid—what would become known as a screen. The screen was placed over a photograph and the screened image rephotographed. Each node where the lines crossed generated a small focal point for the new photograph—lighter areas in the original would produce large dots through these points, darker areas smaller dots. The result was a negative made up of dots where the darkness was reflected in the size of the dots. Although this is a less satisfactory image than a collotype, which handles continuous shading like a real photograph, it was a much more practical way to reproduce an image in print because it used a conventional lithographic plate.

At first, this image was then transferred photographically onto the printing stone using a photographic variant of the greasy resist, but this proved difficult practically and it became more common to print the halftone, and then transfer the image from specially treated paper onto the printing surface. But however the image was transferred, the photograph had been converted into a format that would make it practical to use the efficient lithographic process to reproduce a photographic image. Between the 1880s and 1900, this photolithography would transform printing, taking the illustrated book, magazine, or newspaper from a cottage craft to a mass production industry.

⌘⌘⌘

In the book based on Muybridge's photographs, Stillman claimed that the drawings were used because many of the original photographs were

"imperfect in lights and shades" or "when the subjects were dark col-
ored, were in silhouette." And, he informs the reader, "The outlines [of
the drawings] are quite perfect, and the details in other aspects are
quite unimportant to the study of the movements." This may well have
been a justification of what was more a practical reflection of the difficulty of reproducing photographs, as well as any attempt to reduce the
apparent significance of Muybridge's input.

The book was published in Boston in February 1882 by James R.
Osgood and Company and in London soon after by Trubner & Co.

Despite Muybridge's obvious shock at the Royal Society committee meeting, he may have already been aware of the way that the book
failed to acknowledge his contribution. On March 18, 1882, just 11
days after Muybridge had clearly been ignorant of the book's publication in his letter to Stillman, and four days after his lecture at the Royal
Institution, James Osgood, the owner of the Boston publishing company wrote to Stillman to say:

> Gov. Stanford has authorized us to instruct the London publishers to dis-
> regard the claims of Muybridge (about which we telegraphed you) and we
> have accordingly done so.

It is just possible Osgood was referring to any hypothetical claims
Muybridge might be expected to make, but it is more likely that
Muybridge had seen the London edition and had responded by claim-
ing copyright over the illustrations.

The timing of all this is so tight that it must have involved the
latest in modern communications of the time. Think of the chronol-
ogy. On March 7th, Muybridge did not know that the book had been
published. He must, sometime in the next day or two, have seen the
London edition of the book and got on to the publisher, Trubner, who
then got in touch with Osgood in Boston. Osgood contacted Stanford
in California, who replied authorizing him to instruct the London publisher to disregard Muybridge's claims. And all this before Osgood
wrote to Stillman on the 18th.

Just a few years earlier it would have taken weeks for the message
to get across the Atlantic, and probably even longer to travel from Boston, on the east coast of the United States over to California on the
west. Thanks to the revolution in high-speed communications that had

taken place over the last few years, messages had crossed those huge distances and been acted on at least four times in 11 days.

High-speed communication was nothing new—church bells had been used to raise the alert at the speed of sound, and fire and smoke beacons had flashed messages from hill to hill in ancient times, but these methods were only capable of attracting attention, of letting the recipient know that something was happening—they could not carry the sort of detailed messages that were flying between London and California. Instead, it would have been necessary to fall back to a message carried across the sea by boat, then at the best pace a horse could make—realistically not much more than 10 miles an hour over a sustained period.

In the 1790s, Frenchman Claude Chappe realized that it was possible to make use of the speed of sound by combining an alert with the steady tick of a timepiece. In his original experiments both sender and receiver had a clock. The sender also had a gong, or to be more precise a large French cooking pot. The two clocks were first synchronized with a prearranged series of clangs, produced by hitting the pot hard with a metal implement. Then the message was sent. The two clock faces were each divided into segments, corresponding to the letters of the alphabet. A clang that came when the second hand pointed to "A" meant the letter "A," and so on. It was a slow process, limited to a distance of about half a mile, dependent on wind direction, and highly irritating for anyone in earshot, but it broke the limitation of high-speed, remote communication.

Chappe realized this was only the start. He soon moved across to the faster, less obtrusive medium of light. In his first attempt, he replaced the cooking pot with a board painted black on one side and white on the other. The board rotated on a pivot—the white side was flashed into view as the clock hand passed the appropriate letter. This method increased the range—with a telescope, on a clear day it could work up to 10 miles away—and removed the deafening noise pollution, but was still slow and clumsy. Chappe's final innovation was to provide different visual signals for each letter, so removing the need to wait for the clock hand to make its way round to the right segment.

This move got rid of the clocks entirely, removing an unnecessary layer of complication.

Chappe could have chosen to use the same technique but with several black and white boards that in combination would spell out the letters—in fact, some rival systems would do just this; instead, he produced a device that was based more on the different positions in which human arms could be placed. In Chappe's device two wooden arms were mounted on a tower, and each had an end piece that could be rotated on the arm. Various different combinations of arm and end-piece positions were assigned to the letters of the alphabet, and a practical visual communication device was born. It was called a *télégraphe* from the Greek words for "afar" and "one who writes" at the suggestion of a friend of Chappe's.

Chappe had made his first visual experiments in 1791. By 1794, a string of télégraphe stations had been built between Paris and the city of Lille, a link 130 miles long, covered by 15 télégraphe towers. A good 1,000 télégraphe stations (or semaphores as they became known in English) would be built by the 1830s in countries across the world, linking cities and reinforcing lines of communication that would be important in battles. The télégraphe was expensive to operate, could be devastated by the weather and darkness and was still relatively slow, but for essential messages nothing could beat it.

Even as Chappe's invention reached such dizzying heights, Samuel Morse was beginning to experiment with the use of electricity to carry a message. It would take him 12 years and numerous setbacks before he was able to send the momentous message, "What hath God wrought," on May 24, 1844, along a railroad track from Washington to Baltimore, using a code of short and long electrical impulses that he had devised.

Here was a system that overcame the problems inherent in the télégraphe. It did not depend on visibility, and each character of the message was much quicker and less labor intensive to send. It was still necessary to have repeater stations at regular intervals, but a message could be sent from point to point at remarkable speed. By 1850, in the United States alone there were 12,000 miles of cables for the electric telegraph. In 1866, the other essential step for fast communication between the United States and Europe was put in place—a cable was successfully laid beneath the Atlantic. This remarkable feat of engineering was a reflection of the confidence of the times, given the expensive failures of the first two transatlantic cables, laid in 1857 and 1858, both of which stopped working within hours of first being used. By the 1880s, though, there was nothing remarkable about a message being relayed at high speeds from London to California.

The basis for Muybridge's angry communications with the publisher was the fact that he had successfully taken out a copyright of the hand-printed photographs and the text in the *Attitudes of Animals in Motion* book. Those same photographs formed the basis of and focus of the Stillman book, even though they were mostly in adulterated drawing-based form.

When Stillman's *The Horse in Motion* was originally published, as we might expect today, the London edition was copyrighted in the author's name, while the American edition, rather strangely appeared as the copyright of Leland Stanford. Stillman was later to decide to abandon his rights. He arranged for the United Kingdom copyright to be transferred to Stanford, writing defensively to Osgood:

> I telegraphed you today at the request of Gov. Stanford when I told him of what had been done in the matter of the English copy right of The Horse in Motion. I was satisfied myself that the copy right in my name was wrong and the error should have been corrected and could not be corrected by committing another.
>
> The proprietorship is not in me but in Gov. Stanford alone and it should appear so in the book. In a law suit this may give rise to unpleasant complications.

Stillman was clearly expecting trouble. He went on to try to reassure Osgood, who was probably feeling rather exposed at this point, that he, Stillman, was in the right. In the process he attempted to belittle Muybridge, even using his attempt to get the latest dry plate technology into the book against him:

> With regard to the claims of Muybridge that the illustrations in silhouette are an infringement of his copyright I have this to say that I can swear that they were all taken at the order of Gov. Stanford who paid all the expenses, furnished the apparatus and material and Muybridge furnished me with all the copies from which the plates were executed knowing that they were

to be used for the purpose to which they were to be applied. [Stanford] also furnished him with the magic lanterns and apparatus which he is now using to amuse audiences in England and the money he used to travel and exhibit the movements of animals, and he imposed upon the Governor the idea that he possessed the most delicate chemicals ever used to produce the results when in fact he was far behind the times and processes were in use for years far more delicate and which he did not know of until he went to Europe and he wrote to me when I was in Boston to delay the work until he could take a new set of photographs by the dry process of which he had just learned.

As if he hadn't done enough in criticizing Muybridge's photographic expertise—an expertise that was certainly never called into question by anyone in the United States at the time the photographs were taken, let alone a photographic illiterate like Stillman—the good doctor then went on to emphasize that Muybridge had unfairly taken all the credit for using electricity, and that the photographer had hoped to produce an elaborate, hugely expensive volume rather than the practical book Stillman had written.

It would seem that Stillman wrote this letter in something of a hurry, or in an emotional state. Some of the sentences carry significant emotion, while others were clearly incorrectly written—at one point he says that Muybridge "gave all the credit" for an idea where presumably meant to say "he took all the credit."

Meanwhile, Muybridge had instructed his New York lawyers that he intended to pursue copyright in both the United Kingdom and the United States. Osgood and Stillman made frantic efforts to make sure that the United Kingdom copyright was registered in Stanford's name to make it a clear (and inevitably one-sided) battle between Muybridge and the governor.

Muybridge had, by now, had his crushing encounter with the Royal Society's council. He felt that he had to return to the United States to defend himself against what he saw as the depredation of Stanford and Stillman. It wasn't just a matter of what happened at the society. The publication of the book was resulting in more public speculation about Muybridge's role, and pushing him behind the scenes. When the leading British science journal *Nature* reviewed the Stillman book on April 20, 1882, it did nothing to defend Muybridge's contribution. It began:

> We have received from Messrs. Trubner and Co. a handsome and richly illustrated quarto "The Horse in Motion". . . by J. D. B. Stillman, A. M., M. D. The investigations are executed and published under the auspices of Mr. Leland Stanford of Palo Alto Farm, California.

The item goes on to say that *Nature* hopes to soon provide a review "at some length," but for the moment would make an extract "from Mr. Leland Stanford's preface, which shows the exact part taken by each of those concerned in the investigation." This extract made Stanford the prime mover who employed Muybridge to take the photographs (apparently under Stanford's direction), then employed Stillman to "review the whole subject of the locomotive machinery of the horse." From this it would seem the whole business was Stanford's work with a couple of contractors brought in to deal with minor practical details.

The following week a letter from Muybridge was printed, including the following spirited defense:

> . . . [you] remark [that] "the following extract from Mr. Stanford's preface shows the exact part taken by each of those concerned in the investigation." Will you permit me to say, if the subsequent quoted "extract" from Mr. Stanford's preface is suffered to pass uncontradicted, it will do me a great injustice and irreparable injury.

Muybridge does not deny that Stanford initiated the Palo Alto experiments, but goes on to emphasize just how much the solution to the problem, and hence the technical achievement, was his:

> I invented the means employed, submitted the result to Mr. Stanford, and accomplished the work for his private gratification, without remuneration. I subsequently suggested, invented, and patented the more elaborate system of investigation, Mr. Stanford paying for the necessary disbursements, exclusive of the value of my time or of my personal expenses. I patented the apparatus and copyrighted the resulting photographs for my exclusive benefit. Upon completion of the work, Mr. Stanford presented me with the apparatus. Never having asked or received any payment for the photographs, other than as mentioned, I accepted this as a voluntary gift: the apparatus under my patents being worthless for use by anyone but myself. These are the facts; and on the basis of these I am preparing to assert my rights.

This is Muybridge distancing himself from the various builders and technicians who had operated under his direction. They were employed, admittedly with Stanford's money. Muybridge, however, had been an independent investigator, responsible for the means of achieving the photographs without being employed by Stanford—and this was what made the whole tenor of the Stillman book, from that title page onward, unacceptable.

Muybridge caught the *Republic*, sailing from Liverpool for New York on June 13, 1882, and arriving 12 days later. Back in America, Muybridge launched two legal challenges—against Osgood and Company for violation of his copyright and damage to sales of his own *The Attitudes of Animals in Motion* and against Stanford himself for damages amounting to $50,000.

The legal processes ground on slowly. Although he arrived in New York on Independence Day 1882, it was not until September that his case against Osgood was heard and October that he was to go head-to-head with Stanford.

In the meantime, Muybridge did not let the legal proceedings get in the way of promoting his work now that he was back in the United States and distanced from the scandal at the Royal Society. From shortly after his crossing the Atlantic to the middle of 1883 he made a big hit on the lecture circuit, appearing in both highly respected locations like the Massachusetts Institute of Technology and the National Academy of Design, and rather less renowned venues arranged for him by Kelley's Musical and Literary Bureau in Boston. But the natural enthusiasm he felt for the subject must have been tainted by the awareness of the two lawsuits—a taint that was not to be washed away by the outcome.

The better of the two, from Muybridge's viewpoint, was the dispute with the publisher, Osgood. To the lay eye Muybridge's case seems watertight. He was the accepted author of *The Attitudes of Animals in Motion*, for which he had copyright in his name dating back to 1881. No one doubted his claim to be the originator of the photographs in that book, which were the basis of the illustrations in Stillman's *The Horse in Motion*, which Osgood had published.

Osgood escaped by using the usual publisher's excuse that they were not responsible for breaches of copyright made by the author, or by the person whose name was on the book as copyright owner—in this case, Leland Stanford. Osgood's defense was that he accepted the book in good faith. While very recently some questions have been raised as to whether this is an acceptable stance, at the time, Osgood was considered in the right. The court did not dispute Muybridge's claim to copyright on the images, but the belief was that Osgood had no case to answer, and Muybridge's claim was dismissed without prejudice.

The Stanford "trial" that followed proved to be not a conventional argument between counsel, but rather the presentation of a series of depositions, that is, written testimony, the documentary legal equivalent of appearing in court as a witness. This process is often a slow one, and from Muybridge bringing the suit to the final judgment took four months. That judge ruled in favor of Stanford.

A letter arising from the suit brought out the underlying needling between Muybridge and Stillman, the author of the hated book. Stillman's letter to Alfred Cohen, Stanford's lawyer, makes it plain how he viewed the other man:

> I believe Muybridge to be a very unsafe and unscrupulous man. If he does not wear hay on his horn he does carry a pistol in his pocket and he did shoot a friend in the back and plead insanity.

If Stillman believed what he wrote (inaccurate though the suggestion that Muybridge shot a friend in the back might be), he was taking something of a risk in mocking Muybridge. If this was truly a man who could be so wounded by betrayal as to murder, Stillman was perhaps a good second candidate for assassination. As it was, though, Muybridge made his assault a legal one.

Like the Osgood suit, his failure against Stanford did not undermine Muybridge's copyright, but simply meant that the court believed he had no case for damages against the governor. The action had not cleared Muybridge's name, nor had it prevented the continued sale of the Stillman book, but at least he did still have ownership of the images. And the whole process built up in Muybridge a determination to go further, to show that the work with Stanford was but a stepping-stone on the way to final triumph.

One small consolation Muybridge might have had was a comment on the book made in the scientific journal *Nature* on June 29, 1882, which Muybridge was later to quote. On *The Horse in Motion, As Shown by Instantaneous Photography, With a Study on Animal Mechanics, Founded on Anatomy and the Revelations of the Camera, in which is Demonstrated the Theory of Quadrupedal Locomotion*, by J. D. B. Stillman, A. M., M. D., *Nature* comments:

> The above is the somewhat long title of a large and important work issuing from the well-known Cambridge (U.S.) University Press.
>
> Long as is the title, the name of the principal contributor to the volume is left unrecorded there; though, indeed, even a cursory glance over its contents shows how much indebted is the whole question of the mode of motion in the horse to the elaborate series of investigations of Mr Muybridge.

Another, definite consolation was that the Stillman book simply did not sell well. Reviewers picked out its dull writing style (very different from Muybridge's own) and frequent mistakes. The first edition was badly assembled, with many of the illustrations mislabeled and out of sequence. The line drawings simply didn't have the impact of the photographs they were based on. The second edition was boycotted by many booksellers because there wasn't enough markup for them (Stanford was presumably trying to recoup some of the $6,000 the first edition cost him). Stanford later wrote to Stillman:

> I think a low price [for the book] is best for I would like to hear of some sales and possibly someone may want it if the price is low enough. I have never heard of anybody's buying the book nor have I heard of the book's being sold.

But low price or not, Stanford felt that they should not be too concerned about its sales. He told Stillman that he should not allow the matters to worry him. And if, he commented, people did not buy the book, it was their misfortune as much as it was Stanford and Stillman's.

Muybridge might have been even more determined to prove his expertise and to spite Stanford had he seen a letter that Stanford wrote to Stillman on October 23, 1882. In it Stanford tells of the Muybridge suit. He is adamant that the idea was his (Stanford's), and that Muybridge had frequently said so. But the sentence that would really have set Muybridge's blood boiling is this:

> I think the fame we have given him has turned his head.

From Stanford's viewpoint, Muybridge had earned nothing—he had been "given" fame, and this was something Muybridge would have found hurtful in the extreme. The use of "we" is interesting. Stillman had not been present at the experiments, and had actively excluded Muybridge from any fame. Either Stanford had mentally rewritten history or was using the royal "we."

Stanford's views of exactly what had happened were not limited to private letters. In 1886, the *Boston Herald* published an article on Stanford and his horses. Muybridge is never mentioned by name in the article, but he is subject to venomous ridicule:

> A photographer in California was employed to superintend the matter [of photographing horses in motion], and after these pictures were completed Mr. Stanford went to Europe. The man accompanied him, and a large proportion of the scientific men of France gathered to listen and to see concerning the horse in motion, and then the photographer rather lost his noddle by the attention paid him. He imagined that he conceived the whole matter, and was, so to speak, a discoverer. He ended by putting an injunction on Mr. Stanford's book in America.

While not explicitly stated, the implication is that it was Stanford who the "large proportion of the scientific men of France" gathered to listen to, when in fact he wasn't even present at the event. But by the time this article was published, Stanford was too late. Muybridge had already far surpassed his limited successes at Palo Alto.

TEN

TECHNOLOGY TRIUMPHANT

It has long been regretted as a Misfortune to the Youth of this Province, that we have no ACADEMY, in which they might receive the Accomplishments of a regular Education.

The following Paper of Hints towards forming a Plan for that Purpose, is so far approv'd by some publick-spirited Gentlemen, to whom it has been privately communicated, that they have directed a Number of Copies to be made by the Press, and properly distributed, in order to obtain the Sentiments and Advice of Men of Learning, Understanding, and Experience in these Matters; and have determin'd to use their Interest and best Endeavours, to have the Scheme, when compleated, carried gradually into Execution; in which they have Reason to believe they shall have the hearty Concurrence and Assistance of many who are Wellwishers to their Country.

Those who incline to favour the Design with their Advice, either as to the Parts of Learning to be taught, the Order of Study, the Method of Teaching, the Oeconomy of the School, or any other Matter of Importance to the Success of the Undertaking, are desired to communicate their Sentiments as soon as may be, by Letter directed to B. Franklin, Printer, in Philadelphia.

Introduction to *Proposals Relating to the Education of Youth in Pensilvania [sic]* by Benjamin Franklin, which led to the setting up of the University of Pennsylvania

The lectures that Muybridge gave in America during 1883 after the collapse of his suit against Stanford were to provide more than an income. They were to open up an opportunity to transform his work from the entertaining sequences made for Stanford into a huge systematic study. Fired by the disgust aroused by the Stillman book, Muybridge had for some months been putting about a prospectus for a new series of photographs that would be vastly more sophisticated.

This proposed collection of photographs of "men, horses, dogs, oxen, and other domesticated and wild animals, executing various movements at different rates of speed" would include over 2,000 plates. The theory was that subscribers would be able to mix and match from the images produced by the research (also including everything from "Ladies playing at Lawn Tennis, dancing and other exercises" to "seals and other marine mammals"). Whether it was in response to the prince of Wales's enthusiasm for the entertaining boxers or because of pressure from his contacts in the art world, he intended to concentrate on the movement of the human form. And everything would be vastly superior to the Palo Alto results, using the latest dry plate technology. Of course, Muybridge did not have the funds for this venture—and he would need an awful lot of subscribers if he were to make it happen this way.

Another American benefactor, Fairman Rogers, who had made his money by solving the problem of getting compasses to work effectively on iron ships, had taken an interest for some time in Muybridge's motion photography. When he heard about this remarkable prospectus, Rogers did the best thing he possibly could for Muybridge. Instead of offering to back him, and potentially re-create the problems of the Palo Alto experiments, he spoke to a number of other leading lights of Philadelphia and between them they sold the provost of the University of Pennsylvania, Dr. William Pepper, on the idea of the university's providing Muybridge with the resources to produce a much more in-depth study of animal and human motion. (This Dr. Pepper is unrelated to the soda of the same name, which was invented by an English druggist called Charles Alderton in Waco, Texas, in the 1880s.)

Perhaps the formalized academic route proposed by Fairman Rogers reflected the difference in approach between the more restrained, dignified, moneyed gentry of the East Coast and the ostentatious wealth of the West. Whatever, Pepper was impressed with Muybridge's achievements to date and pulled together a committee of university trustees and benefactors to see whether the proposal could be turned into reality.

The turning point was a lecture at the Philadelphia Academy of Fine Arts and Music in February 1883. A letter Muybridge wrote to the academy detailing the preparation for his lecture, and in particular for using the zoöpraxiscope, illustrates both what was involved in one of his performances and Muybridge's fine eye for detail. He included a drawing of the table he would need prepared, with instructions on how to get it ready:

> An undressed roughly built, strong table 6 feet long, 4 feet wide on top, 6 feet high, with a platform (for men to stand on) 3 feet from the ground, 2 feet wide along each side of the length, and on <u>one</u> end. Some kind of cheap drapery should be hung around the table for its better appearance.

Muybridge then went on to describe the layout of this earliest of movie theaters:

> The table will have to be placed 45 feet from where the screen will be hung, <u>if this distance is convenient it will be best</u>, but if absolutely necessary it can be farther or nearer.
>
> My screen is 16 feet square, the bottom of which—if the floor of the room is level—should be not less than 4 feet from the ground.

Then there was the matter of illumination, and of handling the conventional magic lanterns that would be used for the ordinary slides:

> [S]ee Mr R Munnie of 621 Commerce St, (the agent of the New York Calcium Light Company) and request him to have delivered at the Academy on <u>Monday</u> MORNING, 2 pairs of Cylinders (2 Oxygen, 2 Hydrogen) each pair capable of running full headway for 3 hours, I think I usually have the "25 feet cylinders." Couplings to accompany cylinders. Also that he furnish 2 capable and handy men to attend to the 2 lanterns used, it is necessary both of these men should be accustomed to run a magic lantern and to change the pictures; and it will also be necessary they should be at the Academy on Monday afternoon to receive instructions how to manage the apparatus which is of peculiar construction.

Before electricity was common, these frighteningly dangerous "calcium lights" or limelights were the usual way to illuminate a projector for large-scale audiences. A stick or a ball of lime (calcium oxide) was bathed in the flame of the very inflammable mixture of hydrogen and oxygen—the same explosive mix of gasses used to power space vehicles into orbit. Calcium oxide has a very high melting point and so can be heated all the way to white heat, resulting in an intense, white glow; but it uses an extremely dangerous combination of the easily flammable hydrogen with enough oxygen to produce instant conflagration.

The American chemist Robert Hare first discovered the intense heating power of a hydrogen flame, provided with a stream of oxygen to keep it burning to maximum effect. Around 1800, while still in his teens, Hare used a keg from his father's brewery to set up separate reservoirs of oxygen and hydrogen in the ratio of two of hydrogen to one of oxygen that would enable efficient burning to produce water. The searing flame exceeded the capabilities of all rivals, being able to consume diamond and melt platinum—and to heat calcium oxide to the extent that it glowed with an intense white light. With a melting point of over 2500 Celsius (over 4500 degrees Fahrenheit), lime could stand up to Hare's blowtorch as it reached white heat.

As atoms are heated they give off light. At 1000 Celsius the predominant color is red, but with increasing temperature a mix of the higher end of the color spectrum up to blue comes in. The light will be increasingly white with a bluish tint, making it close in appearance to natural light. This was the effect that was necessary to get the best results from Muybridge's images.

The gas cylinders that Muybridge used were not 25 feet long, but contained a measure of gas based on 25 feet (this could have been the pressure or the volume). Even so, the cylinders could be as much as 8 feet long—the zoöpraxiscope was no pocket projector.

The event went well, as a report in the *Philadelphia Evening Bulletin* of February 9, 1884, showed. After describing how the zoöpraxiscope enabled Muybridge to portray the movements of different animals—and even a man turning somersaults on the back of a horse going full speed—it goes on to remark that "these innovations in the art of the magic lantern displays were received with admiration."

The same report also showed Muybridge to be adept at handling a noisy audience, though it isn't clear from the wording whether he was dealing with booing or cheering. The only unpleasant feature of the evening, we are told, was the presence of a number of students who felt called upon to interrupt the lecturer whenever the university [of Pennsylvania] was mentioned. The caliber of the crowd, according to the reporter, was pretty well shown by the applause that followed a remark by Muybridge that some of the students "had more muscle than brain."

When Muybridge gave his sell-out lecture he would not have been aware of the significance of a small group in the audience from the university authorities, but he was soon to meet the committee and discover what they had in mind. The Rogers-inspired plan was a godsend, and after careful negotiation that lasted through to the end of the summer, Muybridge was able to agree to a scheme that would see funding of $20,000 and University of Pennsylvania buildings and staff given over to making an in-depth photographic study of motion.

This might seem a remarkable step for a university of that period to take, but Pepper was an unusually dynamic provost. At the start of his period of office in 1881, the University of Pennsylvania was generally considered a second-rate establishment, but Pepper worked hard to promote it and expand its facilities, particularly around the natural sciences. He was to start a college of veterinary medicine and introduced at least half a dozen new areas of study, including natural history and philosophy.

Pepper's enthusiasm was infectious. Not only did he contribute from his own pocket but he was also excellent at getting others to come up with funds, and this was the approach he was to take with Muybridge's bold vision. With five others he eventually guaranteed funds of up to $30,000 (over half a million dollars now), so that the requirement of the prospectus was fully covered.

That didn't stop Muybridge from circulating the prospectus, though—the university angels were merely acting as guarantors, and the intention was that Muybridge would raise the cash by selling portfolios of photographs. This was a lot of money to find, but when he first made the arrangement with the university he already had 63 subscribers at $100 each, which would buy them 100 prints of their choice. By the time Muybridge began to take photographs in the spring of 1884 this figure had reached 133 (we know because Muybridge included a list of subscribers in later versions of the prospectus to encourage others to buy in).

In return for the funding, Muybridge had to undertake to make his experiments at the university—hardly a burden, given that he had no premises of his own—and to work under the supervision of a commission of nine university faculty members: Pepper himself; professors of anatomy, veterinary anatomy, physics, engineering, physiology, and civil engineering; the president of the Philadelphia Academy of Fine Arts and Music; and the renowned painter Thomas Eakins.

It is also possible that Muybridge found that the university was a little more worried about appearances than his acquaintances in the West. According to a memorandum in the archives at the University of Pennsylvania, when Muybridge became associated with the university, Pepper told him, "Now you are associated with the university you will have to dress better."

Muybridge apparently asked, "What is the matter with my clothes?"

Pepper, perhaps with an appropriate lift of the eyebrow, came back, "There is a hole in your hat and your hair is sticking up through it."

Muybridge removed the hat and remarked with apparent surprise, "Why it does have a hole in it."

After some discussion, the site for the experiments was agreed on. Muybridge was to use the yard of the department of veterinary medicine. At first sight this was not an ideal location. It certainly did not have the feel of wide-open spaces that Palo Alto provided. The yard, before Muybridge took it over, was an uneven patch of land, mostly rough earth, but with a triangular flowerbed slotted into the acute angle between two wings of the veterinary block.

The buildings on two sides of the yard looked, to the modern eye, like a cross between a Victorian factory and a rural railroad station, but at the time they were newly constructed and considered quite smart. In practice, the restrictions of the site were not to be an excessive problem. There was no intention of capturing trotters at a fast pace here— most of the subjects would be moving in a much more controlled fashion.

One of the first things to go up on the site was a high wooden fence that shut the open side of the yard. The fence was essential, as right from the beginning the intention was to use nude models for some of the sequences, which was of interest both to the medical men and to the painter Eakins, but had to be handled with immense care. Eakins was well aware of this—in fact, he was later sacked from the Pennsylvania Academy of the Fine Arts and Music as a result of a petition initiated by some of his students. He had provided his drawing class with a nude male model, and two female students had collapsed in a faint. Eakins was unrepentant, arguing that it was impossible to properly draw or paint the human form without studying the unclothed body, but he was made painfully aware just how much care was required to make nudity acceptable.

The basic arrangements for the photography were not unlike Palo Alto. There was a track—though only 120 feet long—with a cloth backdrop, and facing it a shed that could house up to 24 cameras. But this endeavor was a more scientific one. Instead of a simple white canvas backdrop, split up by vertical black lines 21 inches apart, this one had a grid of 50-centimeter squares, subdivided into 5-centimeter sections. (These sections were called 2-inch squares in contemporary newspaper reports, but that is likely to have been a simplification for the reader's benefit.)

This use of metric measurements on the grid (it seems rather more likely that they were intended to be 50 centimeters than $19\,^3/_4$ inches) demonstrates the forward thinking of the University of Pennsylvania at that time. The metric system was proposed as far back as the 1500s by Flemish engineer and mathematician Simon Stevin, and was adopted in France in 1795, then dropped by Napoleon and reinstated by the French government in 1840. It took much longer to penetrate the scientific community in the United States and the United Kingdom. It wasn't until 1866 that it was legal to use metric measures in the U.S., and 1875 before there was common agreement on what the basic metric measures should be. And it would be well into the twentieth century before metric measurements—to be precise, the *Système International d'Unités* (SI)—were accepted as the worldwide scientific standard.

The backdrop itself was black, with the grid made out of white string (in some photographs the string has clearly been pulled out of position by an animal's foot or tail). The grid allowed much clearer measurements of the movements and positions than had previously been possible—and provided the characteristic hatched background still seen when Muybridge sequences or modern equivalents are used as graphic images. The black backdrop was particularly well suited to light human skin tones, and could only be used because of the move to the more sensitive dry photographic plates that allowed exposures to be taken with less brightness in the subject. There was also a white backdrop available for darker subjects.

The other big difference here was the understanding that a row of cameras parallel to the track would not always capture everything that was required. In studying muscle movement and details of gait, Muybridge wanted to be able to see more than a side view. As well as the main shed, the new setup allowed for two extra batteries of cameras, one angled to catch a rear view of the subject, the other a front-on view. These batteries were mobile, rather than fixed like the main camera shed, so the angle could be altered.

Surprisingly, given the changeable Philadelphia weather, the track was in the open air. It would have allowed more flexibility if it had been enclosed (and the models would have been less likely to suffer from exposure), but it proved impractical given the size and location of the site, and the need for ambient light, to cover the track over.

As before, Muybridge had an equipment shed immediately adjacent to the camera shed, but he was also provided with more sheltered laboratory space in the capacious basement of the university's Biological Hall. Here the new shutter mechanisms for the batteries of cameras were constructed. They were an enhanced version of the electromagnetic shutters used at Palo Alto. This time, to give more precision, the shutters would be fired with precise timing, rather than by the progress of the subject down the track. A large clockwork motor drove a circular rotor that made electrical connections one after another, firing the shutter mechanisms in sequence. The same connection could fire shutters on both the main battery and the portable ones, thereby taking simultaneous shots from different angles.

This procedure controlled the exposures much more accurately, but did not give precise timings of when each exposure took place with respect to the next, which was essential if useful measurements were to be taken from the photographs. To make it possible to pinpoint the timing of each shot, a tuning fork with a frequency of 100 vibrations a second was electrically stimulated to produce a regular pattern on a moving sheet of paper. Every time a shutter was triggered, the fork was given a surge of power, generating higher peaks on the curve, so the shutter timing could be measured much more accurately than ever before.

Tuning forks had been in use since 1711, when John Shore, an English trumpeter, devised one as a means of tuning his instruments (with a certain humor, he called it a pitch fork). The idea is simple—the frequency with which the two arms of the fork—shaped a little like an elongated horseshoe—vibrate is set by the length of the arms. With careful adjustment these vibrations can be tuned to any desired frequency. The actual frequency of a particular note had first been pinned down by Newton's contemporary and bitter enemy Robert Hooke, who pressed a sheet of cardboard against the teeth of a cog wheel—knowing the number of teeth and the speed at which the wheel was turning, he could match a frequency to the note produced by the vibrating card.

In the musical world, tuning forks are used because the sound produced by the fork varies in pitch according to the frequency of vibration, but Muybridge wanted to use the regular movement of one of the tuning fork arms to draw a rippling line on a moving roll of paper. Normal tuning forks only produce a note for a few seconds, but by the 1880s forks were being produced where one of the prongs of the fork was repeatedly attracted to an electromagnet. The magnet, switched on and off by a simple breaker circuit, provided the impetus to keep the fork in motion, while the natural frequency of the fork fixed the rate at which it vibrated. It was then only necessary to arrange for a surge in current in the magnet when each photograph was taken and the shutter triggering was marked clearly on the regularly spaced series of peaks on the paper roll.

Once the worst of the winter weather was out of the way, Muybridge was ready to begin his huge, if arbitrarily chosen, set of photographic sequences. The human subjects, driven both by the obsession of Thomas Eakins and demands from medical and other subscribers, came first.

In a prospectus published in 1887, when sets of photographs were for sale, Muybridge gave fascinating detail of the models, many of whom were male:

> The greater number of those engaged in walking, running, jumping and other athletic games are students or graduates of the University of Pennsylvania—young men aged from eighteen to twenty-four.

But an attempt was made to go beyond these fit young specimens to give a wider picture of human motion:

> The models 52, 63, 65 and 66 are teachers in their respective professions; 60 is a well drilled member of the State Militia; 51, a well known instructor in art [Thomas Eakins]; 95 an ex-athlete, aged about sixty [Muybridge himself]; 22 a mulatto and professional pugilist; 27, 28 and 29, boys aged thirteen to fifteen.

and so on. Muybridge's reference to himself as an "ex-athlete" was probably not ironic—even though there is little evidence of him engaging in sports, his fieldwork certainly involved plenty of athletic exertion. He also included patients from the university hospital "selected to illustrate abnormal locomotion" and a number of tradesmen of whom "the mechanics are experts in their particular trades, and the laborers are accustomed to the work in which they are represented as being engaged."

As for the women, they were "chosen from all classes of society," including a widow in her mid-thirties, a wide selection of unmarried women aged between 17 and 24 and model number 20, who was "unmarried, and weighs three hundred and forty pounds." Not surprisingly, given the sensibilities of the time, Muybridge found it much harder to get female subjects than male. Because he wanted natural, flowing motions, he dismissed some of the artists' models, who were fine for static work, but incapable of moving effectively. Instead he used anyone from actresses to elegant members of society, provided they were prepared to follow his detailed instructions in the name of science.

Some of the subjects were dressed, but many were either draped in a semitransparent material or entirely nude. Muybridge complained that he had great difficulty persuading mechanics to do their jobs unclothed. Apart from work activities there were a whole series of movements from walking and running to hopping and jumping. Models were photographed on the flat and on slopes, and undertaking everything from ironing to playing sports.

In all, Muybridge would take over 20,000 photographs over the next two years, a process that took a huge amount of time and patience. But lacking the pressure of a taskmaster like Stanford, he was able to work at his own pace, and to his own satisfaction. It appears that this pace was not one that always suited the supervising committee of the university. The artist Thomas P. Anshutz, a student of Eakin's, writing in the summer of 1884 to another student named Wallace, reported that:

> He has not yet done any work with his series lenses and I hear they do not work. The shutters are too clumsy and slow. The University people are dissatisfied with the affair as he cannot give them the results they expected. Which was to photograph the walk of diseased people, paralytics etc. so that by means of the zoopraxiscope (help!!!) they could show their peculiarities to the medical student. This it seems, however, cannot be done even with the best known contrivances. So they would like to fire the whole concern, but they have gone too far to back out.

Anshutz was mistaken in blaming Muybridge for delays in photographing hospital patients, when in fact the hospital authorities were to prove much more of a problem than any technical difficulties. It's also not obvious what evoked his reaction in the letter to the zoöpraxiscope, whether it was just the clumsy name or the thought of using this heavyweight piece of equipment to make demonstrations to students.

For the first few months, not surprisingly, the new batteries of cameras and shutters would not work properly. This was, after all, not an off-the-shelf operation. The shutters and the electromagnetic trigger mechanism that drove them were purpose built and would have taken a lot of fine tuning. In the meanwhile, to have something to show for his time, Muybridge had been taking pictures with a device that was rather like an inverse of the zoöpraxiscope.

In this machine, a pair of contrarotating discs with slots cut in them rotated at high speed in front of a slowly turning photographic plate. When the pair of slots was lined up with the single lens, an exposure was made on the part of the plate that was currently behind the lens. By the time the slots coincided again, the plate would have moved on a little. The result was multiple exposures on a single plate, which Anshutz considered to "show one clear view at every 2 or 3 inches of advance." Sometimes the images ran together, but often the device produced a circle of single images around the edge of the glass plate.

This single-lens camera with a wheel-like shutter was probably based on a device (called the photographic gun) built by the French scientist Marey, which Muybridge had seen on his visit to Paris. He did continue to use it for special cases, for example, when shooting his models with "abnormal locomotion," including a patient with cerebral palsy, another with epilepsy, and one poor soul who was used to demonstrate artificially induced convulsions—but as soon as the main batteries of cameras were working properly they became his primary instrument.

The artificially induced convulsions were produced in a volunteer using electricity after the Blockley Hospital for the Poor, situated near the university, had initially refused to have their patients photographed. The man responsible for inducing the convulsions was Dr. Francis X. Dercum, who worked with Muybridge on the medical case photographs. In a letter written in 1929 to a member of the university faculty, Dercum remembered:

> Such photographs had never been made before. The photographs were the more interesting because they were induced by myself artificially in the person of an artist [sic] model who willingly gave her consent to the experiments.

It's not clear whether the model was aware of the risk of being given powerful enough electric shocks to induce convulsions. Dercum went on to give a word picture of Muybridge himself:

> I knew Mr. Muybridge intimately and well. He frequently dined at my house. He was an Englishman by birth. He possessed strong and regular features and a most attractive personality. In his mental makeup he was more like Thomas A. Edison than any other man whom I ever knew. He called on me after he made his trip to Europe of which he gave me a most interesting account.

The reason the hospital refused consent initially is clear from an item in the *Philadelphia Press,* where it is stated by the hospital committee that their institution would "derive no credit from the work" and "would not reap the benefit of the clinics." While they also expressed a concern that lives would be endangered in moving patients from the hospital, it seems clear that the committee was more concerned about income than the well-being of its patients.

After initially refusing all access, the committee gave permission to take pictures only within the hospital—a very difficult prospect in that dim and dismal building. Finally, however, they conceded that it would be acceptable for photographs to be taken in Muybridge's studio at the university, where the studies were finally produced with no obvious sign of the patients' lives being endangered.

The medical opportunities for making use of Muybridge's photographs were considerable. Take the example of a series of 33 photographs—three sets of 11 taken from different angles—that would appear in the final collection, *Animal Locomotion*, as plate 541. They are described as "Woman with multiple sclerosis, walking." In them a naked young woman is helped along by a nurse, strangely contrasting in her neck to ankle dress. The photographs make it possible to examine much more clearly the adaptations to normal walking posture that are made to cope with a loss of movement in the limbs.

A series from the Pennsylvania experiments—Woman with multiple sclerosis, walking.

These sequences had two benefits. When projected in motion through the zoöpraxiscope they gave medical students who may not have direct access to a sufferer with a particular condition an opportunity to examine the impact of the disease—and at the same time, the individual frames made it possible to study aspects of the motion—whether the gait of the multiple sclerosis victim or the spasms of the artificially induced fit—that simply weren't visible to the naked eye because they occurred too quickly.

The photographs of motion that, according to Anshutz, dissatisfied the university committee were clearly a stopgap. By 1885, a reporter from the *Philadelphia Times* was able to see the full camera batteries in action, with much more impressive results:

> A beautiful young girl, who is a professional artist's model in New York was photographed several hundred times this afternoon. She was of almost faultless figure and her drapery was a loose covering that allowed for muscular action to be distinctly seen. At a word from Mr. Muybridge she left the dressing room at the corner of the enclosure, and walked rapidly to a little platform towards which the thirty-six cameras were pointed.

Sometimes there were 24 cameras in the main battery, but for this session Muybridge was using all three batteries, with 12 cameras in each.

> The photographer held the stick commanding the thirty-six batteries in his hand [the reporter meant 36 cameras], and at another word of command from him the model stepped forward, placed a chair in position, seated herself, crossed one leg over the other, and began fanning herself. These various movements occupied three seconds in execution and every successive muscular movement was photographed from three different standpoints.

Although the reporter makes the process sound continuous, between each group of 12 shots (a different group for each of the actions he describes), the cameras would need reloading. To make this process quicker, instead of having a separate dry plate holder for each camera, there was a huge tray that extended along the whole battery of 12, making it possible to reload the battery in a relatively short time. At another sitting the reporter goes on to say that the model "performed a pirouette and twelve successive impressions of her motion were obtained from three different points of view."

For the reporter this was a matter of some excitement, but very quickly for Muybridge the process must have settled into a routine—but a routine that contributed inexorably to a growing collection of scientific knowledge. Operating this industrial-scale experiment was not a job for a single man. At any one time Muybridge would have up to six students helping him out on everything from loading the cameras' plate holders to setting up the props for a shoot. One of them, Edward R. Grier, later described Muybridge as "a peculiar man," commenting:

> I have seen him with only shirt, pants and shoes on and pants so decrepit that it was not safe for him to go outside the studio.

It would be fair to assume that Muybridge did not take seriously Pepper's advice that he would have to dress more smartly now that he was working at the university.

∞∞

The open air studio at the university was fine for the cooperative human models, who were the subject of the most important part of the work from both a scientific and an artistic viewpoint, but Muybridge also wanted (and had promised in his prospectus) to extend his work to a broader field of animal movement.

The main shed was fixed in place, but because the two extra batteries were movable he could go out on location, and spent two weeks at a racetrack at Belmont on the outskirts of Philadelphia, known as the Gentleman's Driving Park. Here he returned to his first field of success—the horse. Ridden and coach-pulling horses were put through their paces in front of the portable camera batteries and a movable screen. The results made it clear that the combination of dry plates and the improved shutter mechanism allowed for a much better action photograph. Now there could be no accusation of producing mere silhouettes; these were detailed photographs of horses in motion.

The extra space allowed Muybridge to run horses right up to the gallop, capturing each of their unique gaits. Although the horse photographs were a significant advance on the Palo Alto series and much more scientifically precise, the most eye-catching aspect of his animal work resulted from a series of sessions at the Philadelphia Zoological Gardens.

Some early work at the zoo covered creatures that bore a similarity to horses—camels and elk—but Muybridge was then to work through a wide range of birds, from pigeons to an eagle, an ostrich, and a swan. After the aviary was exhausted, the subjects became even more exotic, from tiger to kangaroo. Not surprisingly these attempts caused a lot of interest in the local press. The *Philadelphia Evening Telegraph* and the *Philadelphia Press* both gave a lot of coverage through the silly season in August 1885. The *Evening Telegraph* portrayed Muybridge in action:

> The Professor [*sic*], protected from the sun by an old straw hat, walks about the field like a Western stock farmer. He fixes the slides, gives out the orders, like the mate of a schooner in a gale, and when everything is ready the Professor sits on a small beer keg, holding an electric key in his hand, and calls up the birds.

The *Telegraph* also made clear just how difficult the working conditions were:

> With the sun beating down like a red-hot stove lid, and raising the temperature to something over 100°, work on a white muslin screen is not the pleasantest thing that anyone can wish for. The light is so bright that Mr. Muybridge's assistants wear colored glasses to protect their eyes.

Unfortunately Muybridge was to discover, as has every filmmaker after him who was warned against working with children and animals, that his new subjects had rather different ideas on quite what would happen after he called for action. When the bird-house keeper, a Mr. Murray, opened the box containing a black Antwerp pigeon at one side of the screen backdrop, Muybridge poised ready to activate the shutter mechanism. No bird.

The keeper threw grass at the trap in which the bird was caged. It showed no interest in moving. He threw sticks. Still the bird would not move. The keeper shouted and poked at the animal. It might have been stuffed. It was only when he turned to speak to Muybridge and everyone was distracted that the pigeon shot out of the trap, straight up into the air, not only missing the grid, but also catching the would-be photographer unawares.

Such intransigence was to prove a common occurrence with the birds. Yet Muybridge managed to catch another pigeon in flight in unprecedented detail on his third attempt. This very first successful plate series would establish new facts about the pigeon's frantic, 10-beats-a-second wing movement.

The sheer rate of the wing movement was not known before Muybridge's photographs could be used to analyze it. They also made it possible to identify the source of the characteristic whistling slap of the pigeon's wings as it leaves the ground—the wings coming into contact not, as was then thought, from hitting each other on both top and bottom extremes of the stroke, but at the point they were across the bird's back only. What was less appreciated at the time (though Muybridge would later suggest an opportunity of investigating possible wing shapes and motions for artificial flight) was the way the photographs made it possible to see how wing angles and shapes varied in different aspects of flight with a range of species.

When it came to the big cats, the tiger was suitably cooperative (at least with an element of coercion from his keeper, precariously poised on top of the cage bars with a long stick). But George the lion seemed less impressed. *The Philadelphia Press* of August 22, 1885, sets the scene:

> Photographer Eadweard Muybridge, bronzed and patriarchal, stroked his Rip Van Winkle beard thoughtfully yesterday afternoon, as he stood in the carnivora house, at the Zoo, and looked at the big lion, George, lying in his cage, with his shaggy tawny head between his paws, suspiciously blinking one eye. Mr. Muybridge was very anxious to obtain an instantaneous photograph of a lion in motion, but there were many difficulties in the way.

The lion house was too dark to rely on natural lighting alone, so the white screen had to be lowered at the back of the enclosure. Unfortunately it then covered the entrance to the lion's den, so George and his consort Princess promptly tore down the screen and ripped it to shreds.

This seemed to George to be enough exertion for the day. When a second screen was lowered at the back of the enclosure he totally ignored it and promptly went to sleep. He was awakened in an undignified manner that wouldn't be tolerated on a present-day film set. He was poked in the ribs with a stick until he howled. This irritation was enough to encourage George and Princess to do a bit of classic pacing up and down, which Muybridge managed to capture, though he wasn't ecstatic with the results. He is quoted in the *Philadelphia Times* as saying:

We are laboring under a disadvantage in having to photograph the lions in their cage. I would like the chance to take one in his native jungle, about to pounce on his prey, but as that is impossible, we have to be content with what we can get. We have difficulty with these on account of the strong sunlight which throws the shadow of the cage bars on the lions, causing them to look like tigers.

The day after, the *Press* gives us a typical sample of the popular revulsion with which snakes were treated at the time:

A keeper came out of the snake house at the Zoo yesterday afternoon with a squirming double handful of nastiness. Three young ladies, who had been curiously watching Mr. Muybridge's assistants set up the screen on the asphalt between the snake and the monkey houses and the camera on the grass opposite, arose and vanished squealingly.

The keeper set the pine snake down on the sticky, hot asphalt at the end of the screen furthest from his home. The reptile raised his head twice in the sun, cast a malignant eye at the camera and then uncoiling his five feet of black and white nastiness made rapid and slimy tracks for his lair.

In another example of animal work that would now be considered a totally unacceptable piece of vivisection, Muybridge photographed a living heart beating. One of his student assistants, L. F. Rondinella, later wrote about the process:

[W]e devised a carriage to which a large snapping turtle was strapped on his back, his under shell removed, his heart exposed, and as his carriage was drawn under one of the portable batteries of twelve cameras pointed downward, we made successful series of twelve photographs each, analyzing his heart beats.

Gory though this sounds, it was a rare opportunity to get a better understanding of the exact sequence of compression and dilation in the heartbeat. Although the cardiac cycle, the sequence of movements in the heartbeat, was already known, the photographic sequence made it much easier to examine the movements of the turtle's heart. This was a good choice for the experiment—the turtle heart sustains a stable beat longer than many other creatures in such circumstances. The structure of the turtle's heart is simpler than a mammal's—it has only three chambers to our four, with a single ventricle (the chamber from which blood is pumped out of the heart) to our two. Even so, it provides a good model for learning more about the heart function of all vertebrates, and the Muybridge sequence would provide useful illustration of its cycle.

Hours in the studio and on location were long, and we have little information about how Muybridge lived in this period in Pennsylvania. A glimpse of the eccentric aspect of Muybridge's character coming out is a report from one of his students on the photographer's lunching habits. A favorite was cheese, and Muybridge seemed to positively relish the live maggots that inhabited the cheese he often bought. Echoing some childhood encouragement to eat up his greens even if there were small bugs there, because they would be themselves full of cabbage, Muybridge said that the maggots were just large cheese germs.

Perhaps the best picture of Muybridge in repose at this time comes from a letter written by Erwin F. Faber, an artist who worked with Muybridge on producing zoöpraxiscope discs. According to Faber, the strongest immediate impression of Muybridge was his beard. Even in those times when such bushy beards were more common than the present it was the object of interest; in fact, Faber noted that children

Eadweard Muybridge in 1898 (from *Animals in Motion*).

often stopped the photographer in the street, thinking that he was Santa Claus. Faber describes visiting Muybridge in his room on the second floor of a house on 33rd Street, where he lived "very modestly":

> He had one curious idea—that was that lemons were good for him, and he certainly consumed them by the dozen. He smoked a pipe and occasionally cigars. On one occasion he handed me one to try. I couldn't refuse—and had to take it. It took some courage to go on smoking it, for one or two puffs made me positively dizzy. He watched my face for some expression of approval of the brand—and then—"Faber, what do you suppose I paid for them?"
>
> "Well," I answered, "I shouldn't say less than $5.00 per hundred."
>
> Awfully pleased, Muybridge said, "No, sir, sixty-five cents for fifty."

☙☙

After two summers of photographic work, Muybridge was now largely confined to the basement laboratory. The postcapture part of the work would take nearly as long as the original photography. It was not until a year later, September 1886, that the negatives were prepared for reproduction and the final details of the work were ready to be put into place. By now known as *Animal Locomotion*, this was Muybridge's masterpiece, his equivalent of Newton's *Principia*. The actual components of *Animal Locomotion* started to go out to subscribers in early 1887.

Animal Locomotion was not a book. It was a collection of photographs with a printed introduction and catalogue. The 20,000 originals were condensed down to 781 magnificent 24- by 19-inch plates in the full version, each containing up to 24 photographs—but this sold at a huge $600 (call that $12,000 now). It was mostly bought by large institutions, while the majority of individual subscribers went for the $100 deal, giving them 100 prints of their choice in a leather portfolio. There was also an "author's edition" with 20 plates chosen by Muybridge himself.

Oddly enough, though many universities bought the full $600 set, the University of Pennsylvania did not. When Muybridge moved back to England in 1894 he left a full set of *Animal Locomotion* in the basement of the university library in the hope that the university would decide to buy it, but his hope never seems to have been fulfilled. Perhaps, given the facilities the university had provided, he might reasonably have given them the set.

The laboratory had become nothing short of a photographic factory. A journalist from the *Pennsylvanian* newspaper, visiting Biological Hall in March 1886, reported the various assistants at work on stages of the process of preparation, including one room where "presided over by a pretty young lady with a wealth of golden hair, the multitudinous plates are arranged and classified." The reporter also gave us another thumbnail sketch of Muybridge at this time:

> Mr. Muybridge talks and looks like a philosopher. He is a genial gentleman in the prime of life, and, although his beard is long and white, and hair and eyebrows shaggy, his countenance is ruddy, clean-cut and intellectual. He is intensely enthusiastic in his work, and takes a genuine pleasure in explaining his theories and apparatus to interested friends.

It's not clear whether the readers of the *Pennsylvanian* had much of an idea about what a philosopher looks like, but presumably it included wearing a thoughtful expression.

There were also some thoughtful expressions at a meeting of the project's guarantors in the spring of 1886. There was no suggestion from the agenda that they were still concerned about the results. Muybridge's photographs were by now undoubtedly a triumph, and in the introduction to a monograph on the work produced in 1888, Pepper would write that the result had "fully justified the action of the University, as well as the expenditure of time and money." But the matter of nudity, not for the first or last time was causing concern. By the 1930s, Cole Porter was able to write in his classic song *Anything Goes*:

> In olden days, a glimpse of stocking
> Was looked on as something shocking.
> But now, God knows,
> Anything goes.

But in the 1880s anything didn't go. They were still very much in that "glimpse of stocking" era. Edward Coates, the chairman of Muybridge's guarantors, wrote to Pepper expressing his concerns:

> The human figure series should I think be carefully examined and considered. If the work is to be published at all the usual question as to the study of the nude in Art and Science must be answered Yes. Otherwise the greater number of the [562] series would be excluded. At the same time there are

probably some lines to be drawn with regard to some of the plates. That
there will be objections in some quarters to the publication would seem to
be most likely if not inevitable. Mr. Dickson may be quite right as to the
undesirability of conciliating one enemy but I am inclined to believe this
point worthy of consideration at least.

The Mr. Dickson referred to was Samuel Dickson, one of the rich
benefactors who had pledged to guarantee $5,000 of the costs and a
lawyer by profession. The strange number 562 (in the actual letter
Coates mistakenly wrote 561) isn't a code for a particular series of photographs, but refers to the fact that 562 out of the 781 plates chosen for
final reproduction were of human subjects.

In the 1880s, Muybridge's nude images would have been considered truly scandalous by a part of the population. While stories of pi-
ano legs being covered up to avoid them from raising excitement in
gentlemen are probably apocryphal, it is certainly true that this was, at
least overtly, one of the most prudish periods ever in the public attitude to the display of the human body. That isn't to say that there was
anything unique in this aspect of Muybridge's photographs. As soon as
photography had become capable of mass production there was a
thriving industry in pornographic photographs, which itself was
merely a continuance of a market for titillating paintings that had been
in existence since time immemorial.

In one sense the Victorian sensibilities were probably more realistic than those of other periods in history, when it was pretended that
there could be a distinct line drawn between art and pornography, and
that while pornography was unacceptable, art representing the naked
body was fine, because in some way the mere fact that it was art some-
how removed the erotic connotations. Of course it didn't; there is no
magic wand that prevents great art from being erotic. Twenty years
earlier the great Edouard Manet had shocked the Paris art world with
his painting of his longtime model Victorine Meurent, titled *Olympia*.
According to Antonin Proust (a close friend of Manet's and no relation
to the writer), "If the canvas of the Olympia was not destroyed, it is
only because of the precautions that were taken by the administration." What we now see as a great work of art was then certainly considered unsuitable, horrific, and pornographic (at least in public).

Science was a slightly more acceptable reason for the portrayal of the body—doctors, for example, did have to look at what they were to cure. But the proposal here was not to offer these photographs solely to the medical profession but to put them up for general sale. Thanks to those raked batteries of cameras, many of the sequences included full frontal nudity—something that would not become widely acceptable in photographs, or for that matter in films, outside pornography before the 1960s.

Muybridge and his committee could hardly have been unaware of the potential problem they were creating for themselves before putting all that effort into making the nude sequences. But they had convinced themselves that the combined armor of art and science would prove enough to hold back the sternest critics. When they saw the actual sequences, though, the reality seemed rather different from the concept.

To the modern eye the photographs are largely clinical and unerotic, providing exactly what was intended, an analytical study of the mechanics of human motion, but there can be no doubt that at the time the response of many viewers would be quite different. After all, this was a period when simply picturing a woman sitting astride a horse or bicycle (fully clothed) was considered highly risqué. It is quite possible that there would have been an audience of gentlemen who would happily have added the appropriate Muybridge sequences to their private collections. But should this fact be allowed to stand in the way of the undoubted benefits for artists and scientists alike? To Muybridge's relief, the guarantors and the university decided that it should not.

However, so that there should be no doubt amongst subscribers as to what they were getting themselves into, it was agreed to apply a form of rating system to the catalogue that was to be used by subscribers to choose the plates they would receive. Each photograph would be labeled as "nude," "semi-nude," or "draped." Of course, those going for the full set would not have this preview, but as only a small number of these sets would be sold to institutions, this didn't seem a problem.

The other decision the guarantors had to make was who was going to manufacture the work. It still wasn't practical to print a book containing good-quality photographs, so a hybrid was to be produced—an introduction and catalogue that was printed like a small book, plus a portfolio of loose photographs (that, for an extra fee, could be mounted in albums). According to the frontispiece, the plates in a portfolio would be "enclosed in a strong canvas-lined, full American-Russian Leather Portfolio." The production of the photographic part of the project was outside the experience of a book publisher, so it was handled separately by a photographic printing specialist, the Photogravure Company of New York.

As for the printing of the "book" part of *Animal Locomotion*, there was really no choice to be made. One of the original guarantors was the Philadelphia publisher J. B. Lippincott. He put up $5,000, not purely out of the goodness of his heart but rather on the understanding (admittedly only based on a discussion with Pepper) that the Lippincott company would publish the final work. By the time the project had reached the production stage, J. B. Lippincott had died, but his son, Craige, was aware of the agreement. Because they were printing only the introduction and not the full work, as Lippincott had originally envisaged, it was agreed that the extremely expensive camera lenses used in the experiments, which Lippincott had specifically guaranteed, would be sold and the proceeds given to Craige Lippincott to compensate for the reduced return from their efforts.

During 1887, the first copies of the set of plates were put on display in Philadelphia, Boston, New York, and Washington. Along with the display, Muybridge issued a prospectus and catalogue, which had space at the back to note down the plates you would like in your $100 selection. (Subscribers could, of course, add more at $1 each, or go for the whole set of 781 at $600.) These prospectuses were to be returned not just from those cities but from artists, scientists, and institutions around the world. A useful table was provided for would-be purchasers:

A subscription price of
One hundred dollars for the United States or
Twenty guineas for Great Britain
Or the equivalent of 20 guineas in the gold currency of other countries of Europe
This will be for
Austria 210 Florins
Belgium, France, Italy and Switzerland 525 Francs
Germany 420 Marks
Holland 250 Guilders

Once the pictures were on display and an early version of the introduction produced, newspapers had a chance to review *Animal Locomotion*, and the results were spectacular. It created a storm. The *New York Nation* carried a huge review, beginning with acclaim for Muybridge and the University of Pennsylvania:

> The result of years of labor and of large expenditures of money is at last laid before the public in this magnificent work. The result is one [of] which Mr. Muybridge and the University of Pennsylvania, as well as those concerned either in the actual investigations or in the supply of funds for carrying them on, may well be proud, and the work should belong in every scientific and artistic institution in the country and in the world.

This is unpaid advertising at its best—if Muybridge had written the piece himself (and as far as we are aware he didn't), it couldn't have been more useful.

Next the reviewer gives us a clear picture of the scope of the work:

> The words of the title must be taken in the widest sense. By far the greater number of plates are devoted to the human "animal," while the "locomotion" illustrated includes almost every action of which bones and muscles are capable. It is well known that Mr. Muybridge's investigations began in the attempt to demonstrate the falsity of some commonly accepted and traditional methods of depicting the gaits of the horse. To show how far his work has outgrown this narrow aim, it is only necessary to state that, out of the 781 plates he has now published, only 95 are devoted to the horse, and only 124 to other animals and birds, while the other 562 are devoted to men, women, and children, nude, semi-nude, and draped, walking, running, dancing, getting up and lying down, wrestling, boxing, leaping and playing athletic games—in short, acting before our eyes the animal life of man.

Just in case there was any doubt after that mention of nakedness (and there might be some concern about exactly what those "almost every action of which bones and muscles are capable of" in "the animal life of man" might include) the reviewer now goes on to make sure we are aware that this is done in the best possible (and artistic/scientific) taste:

> Here we have the naked, absolute fact: here, for the first time, human eyes may see just how the human body moves in the performance of its functions, how backs bend and hips balance and muscles strain and swell. This is not an art, but it is a mine of facts of nature that no artist can afford to neglect. How would Signorelli, that enthusiast of movement and anatomy

who drew his dead son naked, or Michael Angelo or Benvenuto, who thought the crupper "a beautiful bone" have revelled in such volumes as these!

"Crupper" originally referred to a leather strap that passed from a saddle under a horse's tail to prevent it slipping off. By Chaucer's time it had come to refer by transference to a horse's hindquarters, and in the seventeenth century became a popular humorous term for the buttocks. As a result of this, the coccyx was sometimes referred to as the crupper bone, and this was probably what the author had in mind. He next gets rather carried away in a flight of artistic fancy:

> How splendid nature is! Here are dancing girls graceful enough to delight the soul of Raphael; athletes with heroic movement that would fire the spirit of Buonarotti; foreshortenings, and flowing contours to satisfy Tintoret: and all with the indisputable stamp of fact, painted for us by the same sun that illumined these glowing chests and cut sharp shadows under the edges of swelling muscles; Thus to see the natural man in his own motion under nature's light is a lesson to humanity of its own glory that the Puritan and the ascetic, the contemner [sic] of the nude and the ignorer of art, would do well to study.

After perhaps cooling down his emotions a little, the reviewer goes on to enthuse more practically over the value of the pictures to artists. He imagined them being inspired and educated by the photographs, though probably not copying their poses directly. In fact, he goes as far as to say:

> [The artist] may even do well to throw away his photographs altogether, the action once understood, and express it by a pose not actually found in any one of them, but conveying better than any of them, taken separately, the total results of the series. For the end of art is not record but expression.

This is not a criticism of the photographs but simply pointing out that the series, while invaluable as a resource to artists, were not art in their own right.

After reflecting on the surprising similarity of some of the pictures to paintings—surprised that the artist in fact got it right a lot of the time—and a brief mention of the animal aspect of *Animal Locomotion,* which clearly provided a lot less interest, the reviewer finishes with another sterling plug for Muybridge.

Although the reviewer was not above a little literary titillation in his wording, there was no suggestion that the pictures might cause offense, or be in any sense distasteful. It looked like the risk taken by the guarantors was paying off. Other reviewers were more sensationalist. The *Philadelphia Press* commented that among the models there was "a monstrous mulatto with the head of a gorilla and the body of a Hercules." And while there was never any outcry at the explicit scenes of nakedness, at least one review in the *Philadelphia Evening Telegraph* warned the potential reader that:

> The subject is a very curious one, if not of enormous importance, while many of the plates are beautiful as well as curious. Some of them, too, it needs to be said, at least approach a line of sensationalism beyond which the dignities disappear, and the propriety of offering some of the plates to subscribers may reasonably be doubted.

If anything, such vague warnings probably improved sales rather than reduced them.

Like many an author after him, Muybridge was well aware of the benefit to the visibility of his book of getting out there and talking about it. In late 1886, Muybridge was putting the cost of the venture at around $40,000, which certainly had not been covered by the subscriptions received by early 1887, even though there was probably up to $25,000 promised before production of the commercial copies began.

From late 1887 onward, Muybridge toured the United States, giving lectures that involved displays from *Animal Locomotion* and using both individual slides and the animations of the zoöpraxiscope. Many of these lectures were free (although there is no doubt that Muybridge hoped to run paying exhibitions later), because they were in essence sales pitches for *Animal Locomotion*.

The need to spread publicity was made clear in a letter that Muybridge wrote to the Reverend Jesse Burk, Pepper's secretary, someone who Muybridge would increasingly rely on as a correspondent and helper when he was away from Pennsylvania:

> It may perhaps surprise you when I say that in the city of Boston I found intelligent men, who not only had no recollection of ever having heard of "<u>Muybridge</u>", but who actually did not know there was a <u>University of Pennsylvania</u> and one of these men subscribed for the work.

> But I did not find a single individual who had not read or heard more or less of the subject of Animal Locomotion being investigated somewhere by somebody.

This letter seems to have been written on the way to Chicago, as it was on letterhead printed "En Route New York & Chicago Ltd.—Pennsylvania Lines—Pullman Vestibuled Train." When he reached Chicago, he was to write to Burke that things went better, as he "did pretty well" with 14 of the $100 subscribers and two complete series at $600 each sold, with more expected to sign up on his return.

While on his journey about the United States, Muybridge was also able to meet up with the man who was arguably the greatest technical genius of the time, Thomas Alva Edison. Two days after his lecture in Orange, New Jersey, Muybridge called at Edison's workshop. This was not a purely social visit. Muybridge would later write in his book *Animals in Motion* that they had discussed the practicality of using his zoöpraxiscope "in association with the phonograph, so as to combine, and reproduce simultaneously, in the presence of an audience, visible actions and audible words." In other words, talking pictures—films with sound—which in practice would not become a reality until the arrival of Al Jolson's *The Jazz Singer* on the scene in 1927. The idea proved unrealistic, Muybridge comments, because "at that time the phonograph had not been adapted to reach the ears of a large audience, so the scheme was temporarily abandoned."

The problem was that the phonograph was a purely mechanical device. The earliest machines used a rotating cylinder covered in tin foil, which was soon replaced with a hard wax. The recording was made by speaking (or shouting) into a tube that led to a diaphragm. The compression and rarefaction of the air vibrated the diaphragm, which in turn vibrated a needle resting on the rotating cylinder, leaving a trail of indentations of varying depth.

On playback, a second lighter and less damaging needle fed the vibrations back to a diaphragm and re-created the sound. The result was inevitably a sensation. When Edison took an early model along to the *Scientific American* office they responded with enthusiasm:

> Mr. Thomas A. Edison recently came into this office, placed a little machine on our desk, turned a crank, and the machine inquired as to our

health, asked how we liked the phonograph, informed us that it was very well, and bid us a cordial good night.

But the large horn used to pick up the sound from the playback diaphragm and push it out into the air limited the phonograph's ability to amplify the sound. At best these devices could fill a room—they simply weren't capable of playing to the sort of mass audience that the zoöpraxiscope could entertain with its projected pictures. Talking pictures wouldn't become practical until electric amplification, by now used with Berliner's flat gramophone discs that had replaced Edison cylinders from around 1910. Even so, looking beyond what was practical at the moment, Muybridge saw a future for this technology:

> [I]n the not too distant future, instruments will be constructed that will not only reproduce visible actions simultaneously with audible words, but an entire opera, with the gestures, facial expressions, and songs of the performers, with all the accompanying music, will be recorded and reproduced by an apparatus combining the principles of the zoöpraxiscope and the phonograph, for the instruction and entertainment of an audience long after the original participants have passed away.

The early moving pictures with sound had no direct link between the picture and the sound track; it was simply a matter of starting the disc playing on cue and the two ran entirely separately. It seems eminently possible that this could have been achieved with a combination of the zoöpraxiscope and a phonograph, though both were much too limited in duration (the early phonograph cylinders lasted just two minutes, while zoöpraxiscope discs were limited to a few seconds) to ever be able to capture an entire opera as Muybridge envisaged.

This method of providing motion picture sound with separate audio discs was just as much a technological dead end as the zoöpraxiscope was for the visual side of moving pictures, but it would provide the stimulus to find a more practical way to link sound to the moving picture, which would eventually come by recording the sound track directly onto the celluloid as a visual representation of the sound (amazingly, this feat was first attempted for sound alone in 1901). Although this approach brought its own technical problems—the pictures had to pass one at a time through the projector, being jerked into place a frame at a time, while the sound had to pass through smoothly through a reader without breaks—it overcame the nightmare complexities for the operator of running separate sound discs and projectors.

Did Muybridge inspire Edison to create a phonograph with pictures? While a statement by Edison at the time confirmed that the visit took place, it explicitly denied that any such subject had been discussed. Yet it wasn't long after that visit that Edison did indeed experiment with combining sound with the moving image using a cinematograph device of his own.

We know that Edison was not above trying to push out a competitor, even one who beat him to an idea. In 1879 he produced an electric light bulb, claiming to be first on the scene, despite the fact that the English scientist Sir Joseph William Swan had demonstrated a bulb, like Edison's based on a carbon filament, nearly eight months earlier. Swan, much less of a businessman, hadn't bothered with the level of patent applications that Edison had. Nor had he the same cutthroat commercial sense. Edison's reaction to the news of Swan's invention was to launch a patent infringement prosecution.

Patent law often seems to favor the commercially strong rather than the most original thinker, but in this case Swan's earlier invention was recognized by the court and Edison failed. As part of the court settlement, Edison was obliged to recognize Swan's independent and earlier invention and to set up a joint company, the Edison and Swan United Electric Light Company, to exploit the incandescent bulb. If this tactic is typical of Edison's approach, it would not be surprising if he had learned more from Muybridge than he ever admitted.

It is interesting that Edison filed a caveat with the Patent Office— effectively a warning to others that he had already thought of something—on October 17, 1888, describing in very broad terms a device that would "do for the eye what the phonograph does for the ear." This was only months after Muybridge's visit. Whether or not his ideas were inspired by his discussion with Muybridge, he was certainly one of Muybridge's subscribers, who sent back his request for a $100 set with the note, "Tell him to make the selection for me."

Edison later wrote to Muybridge to say that he had

> built a little instrument which I call a kinetograph with a nickel & slot but
> I am very doubtful if there is any commercial feature in it & fear they will

not earn their cost. These Zootropic devices are of too sentimental a character to get the public to invest in.

This could have been a bit of deviousness on Edison's part to make sure that Muybridge made no attempt to pursue him financially. The kinetograph, more meaningfully renamed the kinetoscope, was to be a great success, even though it was, in effect, a peep show with kinetoscope parlors, an early form of amusement arcade, set up across America and Europe for the public to drop in their change and watch.

Although the very earliest experiments in the Edison laboratory (the kinetoscope was largely the work of Edison's assistant William Dickson) were suspiciously close to the zoöpraxiscope, with a series of photographs fixed around a phonograph cylinder, Dickson was soon to be inspired by Carbutt and Eastman's newly available roll films. These films were an extension of dry plates, coated onto a long strip of celluloid instead of a glass plate. In the kinetoscope the film ran horizontally from spool to spool, with a rotating shutter masking the movement of the film. By the time they became a commercial success in the early 1890s they had a continuous loop of film passing back and forth over a set of rollers, so that viewer after viewer could watch the movie without rewinding.

Popular though they were, kinetoscopes proved short lived, dying out once Muybridge's approach of projecting moving pictures was married with the new roll-film technology.

Edison was more buoyant on the matter of moving pictures in a letter he wrote that was published in the June 1894 *Century Magazine*. Written before his letter to Muybridge, Edison here seems much more positive about the kinetoscope and future developments—increasing the chance that he was intentionally playing down the business value to Muybridge. Amusingly, Muybridge's description of the future of talking pictures in *Animals in Motion*, published a good five years later than Edison's letter, seems to have been lifted wholesale from Edison.

> In the year 1887 it occurred to me that it was possible to devise an instrument which should do for the eye what the phonograph does for the ear, and that, by a combination of the two, all motion and sound could be recorded and reproduced simultaneously. This idea, the germ of which came from a little toy called the zoetrope, and the work of Muybridge and Marie [*sic*—presumably Marey], and others, has now been accomplished,

so that every change of facial expression can be recorded and reproduced lifesize.

The Kinetoscope is only a small model, illustrating the present state of progress, but with each succeeding month new possibilities are brought into view. I believe that in coming years by my own work, and that of Dickson, Muybridge, Marie and others who will doubtless enter the field, that Grand Opera can be given at the Metropolitan Opera House in New York, without any material change from the original, and with artists and musicians long since dead.

Compare that with Muybridge's "an entire opera, with the gestures, facial expressions, and songs of the performers, with all the accompanying music . . . for the instruction and entertainment of an audience long after the original participants have passed away." Edison acknowledges the contribution Muybridge made and despite denying discussing the subject on Muybridge's visit, does claim to have had the idea of talking pictures around the same time as that visit.

Perhaps the final comment on Edison's involvement should come from a letter written in 1930 by Edison's personal assistant. In it he is responding to a Mr. Armat, who had angrily pointed out in a letter in the *New York Sun* that played down Edison's involvement in the development of moving pictures in favor of another inventor (not Muybridge):

Like yourself, I am filled with righteous indignation when these cheap blatherskites seek to rob Mr. Edison of the honor that is due to him as the inventor of the moving picture art as we know it today. Unfortunately, the public does not know that they are only windbags that would flatten out on being pierced by the truth.

For Muybridge, then, Edison was not to prove the potential partner in developing the zoöpraxiscope that he had hoped for. Muybridge carried on with his successful tour of America while planning to take on the world.

ELEVEN

MAGICIAN OF THE SILVER SCREEN

ANNOUNCEMENT: By invitation of the FINE ARTS COMMISSION of the World's Columbian Exposition, MR EADWEARD MUYBRIDGE will give at intervals, from May to October 1893, in the Zoopraxographical Hall of the Exposition, a series of lectures on the science of Animal Locomotion, especially in relation to Design in Art. . . .

Although it is probable that the present series of lectures may not be unworthy of the attention of Philosophers, they will be free from technicalities and adapted not merely for the instruction but also for the entertainment of popular and juvenile audiences.

From a flyer for the World's Columbian Exposition,
Chicago, 1893

Through much of 1888 Muybridge continued to trek back and forth across America lecturing and raising awareness of *Animal Locomotion.* This tour seems, in Muybridge's eyes, to have been a precursor to a triumphant return to Europe. The snub from the Royal Society in London was still painful, and he wanted to make sure that his true scientific worth was made very plain to those worthies who had rejected him. Being the talk of the United States was very fine indeed, but he wanted to be recognized in Europe, too.

From 1889 until 1891, Muybridge spent much of his time on the road, touring British cities from Bristol to Edinburgh, from Dublin to Manchester. His lecture on "The Science of Animal Locomotion in Its Relation to Design in Art" premiered at the Royal Institution. Shortly after, he was vindicated with an invitation to present the same lecture at the Royal Society.

The society never formally withdrew its rejection of Muybridge's earlier monograph, but by now it accepted Muybridge as the author of the Palo Alto work as well as the far superior photographs from the University of Pennsylvania. Muybridge left London under a cloud after the debacle at the Royal Society eight years before, but now he took the country by storm, appearing appropriately like a stern portrait of God on the cover picture of the popular *Illustrated London News* for May 25, 1889. Not only did he have successful engagements in private venues and at the two great royal scientific bodies, he spoke once again at the Royal Academy and added many other institutions, universities, and major schools to his tour.

As well as promoting sales of *Animal Locomotion*, Muybridge was making money from his appearances, and with typical showmanship made sure that his potential audiences knew that this was not too highbrow an event. In a letter he wrote when looking for new venues in continental Europe, he comments:

> It may perhaps not be irrelevant for me to say that the apparently abstruse subject of Animal Locomotion is, in these lectures, treated in a <u>popular</u> manner, and the demonstrations of the Movements of Animals and Birds are as attractive to the Child as to the Savant, or as the Saturday Review expresses it, "They appeal to every class and condition of humanity" being as well suited for exhibition in the drawing room as in the theatre of a college.

In early 1891, now in his sixty-first year, he moved on to Germany, Italy, and France with similarly successful results. The whole world was eager to see the Muybridge road show. On July 15, 1891, Muybridge wrote to Jesse Burk at the University of Pennsylvania, asking to be sent 400 more copies of the *Animal Locomotion* catalogues (he helpfully pointed out that they were "in the room upstairs where the glass plates are stored"), as he was rapidly running out of stock.

But despite the enthusiasm of the audiences, Muybridge felt that he was living in a bubble of unreality. It was all very well to undertake these tours, but there had to be something else, something more real, to follow. Muybridge wanted to perform further research, and this time to have a proper academic footing—to get his hands on the professorship that some of the newspapers had already awarded him. It is also possible he found the constant traveling a chore. Edward T. Reichert, a physiologist who worked with Muybridge on the University of Pennsylvania photographs (particularly the series exposing the poor snapping turtle's heart) met up with him on one of his lecture tours:

> He appeared to be oppressed even morose, said very little about his experiences in endeavoring to dispose of his pictures, and said nothing about a resumption of his work, or of any plan for the immediate future. . . . He surely was a strange character, but withal very likeable when you knew him well.

This "strangeness" Reichert elsewhere described as eccentricity:

> I became intimately acquainted with the most eccentric man I ever knew intimately. He was very much of a recluse, so that very few got to know the man, and likely no one ever learned hidden secrets that must have radically influenced his life.

It seems likely that Pennsylvania felt it had done its bit already, and did not want to contribute any more funds for his work. Muybridge was to write to the president of the Leland Stanford Junior University—the academic institution that had been built beside the site of his original experiments at Palo Alto.

Although usually referred to simply as Stanford, the now great university is still technically the Leland Stanford Junior University. At the time it was very new, only opening the year before in October 1891, under its first president, one David Starr Jordan, who had been recommended to Stanford by Andrew White, the president of Cornell. Jordan owed his position to Stanford and was unlikely to have been unaware of the ill feeling between Stanford and his now-famous photographer. If that were the case, he would have raised an eyebrow over the letter that Muybridge now sent him. Muybridge begins by saying that though they weren't acquainted, he imagines Jordan knew something of his work and its beginnings for Stanford. He goes on:

> I am now happy to say that after several years of privations, and laborious exertions, I have recovered the position—at least in reputation—from which I was displaced by the publication of "The Horse in Motion by J D B Stillman"; and have completed under the auspices of this University [he wrote on University of Pennsylvania notepaper] a comparatively exhaustive investigation of Animal Movements, and I avail myself of the opportunity to send you two pamphlets on the subject which I think will interest you. I have recently seen both Mr. and Mrs. Stanford, and was very much gratified with the renewed assurance by Mrs. Stanford of her belief that the

interest and attention I succeeded in obtaining for my work by the Senator, during a period of great mental anxiety was mainly instrumental in saving his life.

Muybridge couldn't resist getting a dig in about the Stillman book, but then he made that rather remarkable claim about what Mrs. Stanford said. He does seem to have got on better with Jane Stanford than her husband, and it is possible that they had an equitable meeting, though it seems less likely that Stanford himself was particularly pleased to see the man who had tried to extract $50,000 from him in the courts. However, the claim to have been instrumental in saving Stanford's life is an odd one. The only obvious period of mental anxiety was when Stanford's only son Leland Jr., after whom the university was named, died of typhoid in March 1884. But this was long after any time that Muybridge's work could be said to have provided "interest and attention" for the governor. His letter to Jordan continued:

> I am now being urged by many men, eminent in various branches of Science—among them are Professors Helmholtz, S P Langley, Ray Lankester & Sir Wm Thomson and Sir John Lubbock and Mr Edison—to undertake an investigation of the flight of insects; they being impressed with the belief that a comprehensive knowledge of the subject will be of much interest and value in many ways, and materially assist in the solution of the problem of aerial navigation.

> With this object in view I have devised an apparatus which will be capable of photographing a dozen or more consecutive phases of a single vibration of the wing of an Insect in flight; even assuming that the number of vibrations exceed 500 a second, which is far in excess of what Professor Helmholtz believes to be possible with a common house fly.

We have to wonder why these many eminent men were not suggesting that Muybridge undertake the research under the auspices of their own institutions. His new topic of interest, insects, was clearly less likely to excite either artists or the popular audience, but Muybridge—as ever, one with an eye to the practical—was careful to highlight that this wasn't just about satisfying academic curiosity but could also assist with the "solution of the problem of aerial navigation"—human flight.

Muybridge probably had flies in mind as a subject of study because unlike the majority of insects, they have only a single pair of wings. Had he undertaken the experiment and been able to have captured and analyzed the motion of insect wings, Muybridge would certainly have revealed a lot that was not known at the time, though it is doubtful how much would have helped with the business of aerial navigation. It's true that many early attempts at human flight did involve flapping wings—it seemed natural enough when the vast majority of birds and insects flew this way—but it would never be a practical way to achieve manned flight.

Not only is it very difficult to generate enough lift by flapping wings to get anything other than a very light payload off the ground, if Muybridge had managed to photograph a fly in flight he would have realized just how complex the fly's flight technique is. A fly's wings change shape and go through complex movements, producing an unstable and unpredictable airflow. In some cases, for instance, the insect brings its wings together and then peels them apart from the front, sending a cup of air flowing back from the wing surface. It may well be the vortices generated by the two edges of the wing that provide the lift that keeps the fly in the air, rather than the more obvious "wind" from the flapping. In an even more complex interaction, some of the lift appears to come from the interplay between the vortices and the movement of the wing slightly later, a process called "wake capture."

Although today's computerized technology is almost at the point that it could change the shape of a wing during flight to give extra maneuverability, using part of the fly's technique without the flapping, such complexity would have been impossible with the technology of Muybridge's day. If he had gone through with this experiment, Muybridge would probably have realized that a different model with less complex movements (like a gliding hawk) was more appropriate. He would also have shown that the estimates of the day on insect wing speeds weren't too bad—a housefly's wings can produce 200-300 beats per second.

After speculating on these interesting scientific possibilities in his letter to the president of Stanford University, Muybridge goes on, in a nice piece of promotional work, to point out how his work had increased the visibility of the University of Pennsylvania, and was likely to do so equally for Jordan's newly founded institution. Finally, he makes it clear that he isn't trying to pull strings:

I have made no mention of this matter in any way to Mr. Stanford, prefer-
ring to first communicate with you, and to afford you an opportunity of
proposing it to him should you feel so inclined. I believe the work can be
more successfully carried out in California than anywhere else, and it would
be some satisfaction to me, and it might be also to Mr. Stanford that the
investigation of this subject of Motion should be completed, where it was
commenced.

In one respect at least, this wasn't just an attempt to play on the
past. California provided an excellent environment for the photographer. It possessed both dryness and bright, even light, an advantage
that Muybridge was probably ruefully aware of as he attempted to work
under the much more variable weather of Pennsylvania. For the same
reason, Hollywood would eventually attract the early movie makers,
who were as reliant on natural light as was Muybridge. Palo Alto would
have made an excellent location for further experiments.

Even so, it is hard to believe that Muybridge was hoping that
Stanford, who Jordan would undoubtedly have consulted, was ready to
forgive and forget. From the apparent lack of positive response,
Stanford had no interest in the proposal. Whether or not his claim to
have recently seen the Stanfords was true, Muybridge was to have no
more effective dealings with his one-time sponsor.

Muybridge's last communication with the senator (Stanford was
elected to the U.S. Senate in 1885) was a brusque request in August
1892 for the return of some boxes of equipment left at Palo Alto in
1881. He never seems to have gotten a response, though Stanford was
by then already an ill man (he would eventually die in June of 1893),
and quite probably thought the matter unworthy of his consideration—if the note ever got through to him.

☜☞

In late 1892, after successfully taking on Europe, and while still won-
dering how to get a university post, Muybridge was putting together an
itinerary for a grand tour of Japan, Australia, and India. However, his
plans were thrown into disarray. For once though, the problem was a
positive one: an offer he could not refuse.

For the last few years, preparations had been underway on the outskirts of Chicago for a very special world's fair. These large-scale events,

descendants of Prince Albert's Great Exhibition, had become a popular distraction, but this was to be a dramatic celebration of the passage of 400 years since 1492, marking Columbus's arrival in the New World.

The offer that Muybridge made very little effort to refuse was to appear at this fair, the World's Columbian Exposition. And not just to give a single lecture. He was to have his very own hall in which to deliver talks on animal locomotion, accompanied naturally enough by displays from the zoöpraxiscope.

This was an opportunity that appealed to the showman, the educator, and the entrepreneur in Muybridge. He would have the opportunity to entertain and to educate the masses—masses on a scale that he had never before managed to reach. And though he was to appear at the request of the Fine Arts Commission, this was nonetheless a commercial venture. Muybridge was expected to pay for his own building to be erected on the exhibition site. But in return he could charge admission and sell photographs and books to the visitors. It didn't take long for Muybridge to come back with a "yes."

There's no doubt that Muybridge liked to have the comfort of an external backer, but equally this proves that even at the age of 63 he was prepared to risk his own capital. He was liable for $6,000 for the building, and he was taking the risk that he would recoup that and more.

His building was to be no sideshow booth. The grandly, if unpronounceably titled, Zoopraxographical Hall had the appearance of a majestic post office, about 15 meters (50 feet) wide by 24 meters (80 feet) long. Although it was constructed of wood and iron, it looked as if it were built from slabs of white stone. The façade was covered in a material called staff, a mix of plaster of paris and hemp that was used to face most of the buildings at the exposition, giving the main site the nickname "White City." The hall was fronted with a wide portico, held up by two ornate classical columns. On the roof at the front of the building three tall flagpoles carried banners when a performance was about to be given.

In terms of the exposition, this was a midsize building; there were plenty larger and a stunning array of pseudo-European and -Eastern magnificence fronted on gondola-frequented canals. But the fact remains that this establishment, dedicated to the zoöpraxiscope and Muybridge's lectures, was an impressive setting. It was also, without doubt, the first ever commercial movie theater. Here, for the first time in the world, was a building purposely built to show moving pictures to a paying public. And had he done nothing else, this alone would establish Muybridge's position as one of the founding fathers of the movies.

There was inevitably a conflict of interest between the desire to make his contribution to the exposition educational and to make it a commercial success. The general public, like the Prince of Wales before them, were more interested in seeing boxers pummeling each other (or scantily clad women parading up and down) than in studying the detailed leg sequences of a horse. Muybridge had to bring in the crowds, but at the same time he wanted to be taken seriously as an academic, a contributor to both art and science.

The advertisement for his show by the Fine Arts Commission displays the real dilemma facing Muybridge and his supporters:

> By invitation of the FINE ARTS COMMISSION of the World's Columbian Exhibition, Mr. EADWEARD MUYBRIDGE will give at intervals, from May to October, 1893, in the Zoopraxographical Hall of the Exposition, a series of Lectures on the Science of Animal Locomotion, especially in relation to Design in Art.

> These lectures will be given under the auspices of the United States Government BOARD OF EDUCATION, and will be based on the electro-photographic investigation of the movements of animals, made by Mr. Muybridge for the UNIVERSITY OF PENNSYLVANIA.

> Although it is probable that the present series of Lectures many not be unworthy of the attention of the Philosopher, they will be free from technicalities, and adapted not merely for the instruction, but also for the entertainment of popular and juvenile audiences.

Not exactly gripping. It's easy to imagine what could have been done. See the amazing moving pictures! Witness a lion, greater than life-size, prowling the screen. (Ladies of a delicate disposition may find the experience distressing.) See, re-created through the magic of electric photography, real athletes and beautiful young women dance, jump, and run. Marvel at the boxer's art motion by motion, as real as seen from any ringside seat. And so on. That would certainly have worked better against the competition Muybridge faced. The World's Columbian Exposition was a cross between a trade fair and Disneyland—there were plenty of exciting-sounding attractions. Although there was a huge swathe of the exposition that was undoubtedly worthy and educational—with great machinery and agriculture halls, buildings for each state and many foreign countries—there was a distinct pleasure arm, a wide boulevard, well over half a mile long, leading off the main site and called "Midway Plaisance." (The main drag in American carnivals is often called the midway in memory of this, the first true ancestor of Coney Island, theme parks, and other such pleasure grounds.) It was here that Muybridge's Zoopraxographical Hall was sited, but surrounded by many more exhibits that were less coy about trumpeting their excitement value.

Imagine strolling down the Midway Plaisance amid crowds of people, all in their Sunday best. There is an atmosphere of carnival combined with the thrill of the latest technology and the exotic taste of faraway places. There are exciting new sounds and smells to investigate all around. Yes, you could spend a mere 25 cents to enter Muybridge's hall, but to your left is a huge Ferris wheel, the first ever Ferris wheel and a giant at that, towering 264 feet (80 meters) over the midway. At 60 cents this was one of the more expensive attractions, but it was a unique experience.

A little further along is a dramatic living panorama of the volcano Kilauea at 50 cents. There's an ice railway, a bargain 10 cents per trip, a Lapland village (25 cents), and a tethered balloon ride (at a massive $2 each). Not to mention the heady free attraction of a French cider press. To your right there's a Japanese bazaar, also free, an animal show at 50 cents to $1, a Venetian glass works (25 cents), and a scale model of St. Peter's in Rome (also 25 cents). And then straight across the street, temptingly close and only 15 cents for entry, is "A Street in Cairo"—though you had to pay more to see a show.

A Street in Cairo was a real crowd magnet, according to later reports the most popular exhibit at the exposition, with more than two and half million visitors over the six-month running of the event. The street included a bazaar and the chance to take camel and donkey rides, but the most talked-about feature was the exotic *danse de ventre*, later called belly dancing or hootchy-kootchy dancing. This performance proved quite a shock to the locals. A letter to the *Chicago Tribune*, which could have been genuine (though it wasn't unusual for letters of complaint to be used to publicize an attraction), demonstrated a typical reaction:

> It is a depraved and immoral exhibition. It may well be styled an outrage to allow such an exhibition and rate it under the head of dancing. It is no other than a slightly modified version of the orgy practiced and known in Spain as the "Chica", which was carried into that country by the Moors in the eleventh century, and which was finally forbidden by edict.

The most famous character said to have appeared in the show was Little Egypt, a name that carried a frisson of risk at the time. According to legend, Little Egypt saved the exposition from bankruptcy, killed Mark Twain by inducing a heart attack when he saw her perform, and was filmed by Edison with his early motion picture device, the kinetograph. In fact, none of these seem to be true. While there certainly were exotic Eastern dancers at the exposition, there is no contemporary record of Little Egypt appearing. Sol Bloom, the 23-year-old entrepreneur who had accidentally become entertainments director of the exposition very specifically denied that Little Egypt ever performed there. (Bloom had not wanted to come to Chicago from San Francisco and so had asked the ludicrously high wage of $1,000 a week, only to have his request accepted.)

There is considerable mystery as to who Little Egypt actually was. The name was later widely used by several Middle Eastern dancers, the earliest certain example being one Ashea Wabe, who became notorious when appearing at a raided party held by the Barnum family in 1896 in New York. The name was also later used by Farida Mazar Spyropoulos, who probably was at the Columbian Exposition, but certainly wasn't a star or using that name then. Whoever was on stage, the dramatic performances across the street put Muybridge's effort in the shade. With all the exotic exhibits to distract you, it would have been very easy to have passed by the worthy-sounding Zoopraxographical Hall.

There are no detailed revenue figures from Muybridge's stay at the Columbian Exposition, but it is clear from later maps of the site that his original six-month contract was not extended, so he could not have been raking it in. Even the book that Muybridge wrote to sell to visitors to the Zoopraxographical Hall, called *Descriptive Zoopraxography*, displays the same difficulty of trying to please too many different audiences. It sounds too much like a dull academic title, rather than an exciting souvenir of your first sight of moving pictures. Yet there is no doubt from the subtitle, "the Science of Animal Locomotion made Popular," that it was intended to reach a more general public than his earlier work.

⸿⸿

It is quite probable that Muybridge, at 63 no longer as young and lively as he once was, found the effort of presenting at the exposition day after day left him bone-achingly weary. Even if it had been a financial triumph, he may well have not continued beyond October 1893, when his contract expired. As it was, it seems that he had simply had enough. He no longer turned his mind to exotic tours of the Far East or getting a research place at a university. It was, he seemed to have decided, time to go home. And for all the real and deep affection he felt for the United States, home was still England.

When Muybridge sailed back to Europe in 1894, he was not leaving America for the last time, yet from now on he would be based in England. Initially, he stayed at what had been his grandmother's home in Hampton Wick, then in a boarding house called the Chestnuts in Kingston High Street, not far from his first home, before settling in his cousin's villa at 2 Liverpool Road back in Kingston-upon-Thames. He had come full circle.

Though perhaps he had slowed down, Muybridge had certainly not retired. He had a new book in mind, and had every intention of making use of the well-polished lectures he had developed for the exposition on a more leisurely schedule now that he was back in England.

The book was also inspired by his experience of dealing with a popular audience. Muybridge's intention was to get across the key points that *Animal Locomotion* illustrated but to do it in a way (and at a price) that would interest the general public. When it came to dealing with possible publishers of the book, at the time called *The Motion of the Horse and Other Animals, in Nature and in Art,* Muybridge found there was a clash between his entrepreneurial style and that of a publisher.

Up to now, most of Muybridge's experiences with publishing had been private editions where a sponsor would pay for the book or collection of plates to be published and then the return would be reaped directly by Muybridge and his associates. This was to be a more conventional publishing deal, and Muybridge did not like the figures that were being bandied about. He moaned about it in a letter to the Reverend Jesse Burk, back at the University of Pennsylvania.

Now that Muybridge was in England he tended to use Burk as a representative in the United States, often asking him to go and hunt out a negative or book or piece of equipment from storage and send it off somewhere. This time, though, Burk was just a sounding board for his irritation.

The letter, written on August 5, 1895, starts by clearly responding to a health complaint from Burk in a previous letter:

> I regret very much to hear that the rheumatism has seized upon your hands so badly that you don't feel like writing; now I am going to make a suggestion—not in view however of expectation or wish that you should devote any more of your valuable time in writing to me, than you consider is in my interest—advisably and upon <u>important</u> matters. In your letters why do you <u>write</u> any more than your name? Surely the University ought to furnish its Secty not only with a machine, but also with a stenographic type-writer. I have often wondered how you manage to get through all your correspondence and thought it about time for you to make a change in your method.

A "type-writer" at this time was the person who used the "machine" and stenography ("narrow writing" in the Greek) was shorthand, so Muybridge was quite sensibly suggesting that the university should have a shorthand typist at his disposal. Typewriters had been commercially available for about 20 years, and were becoming much more common, though they came in all shapes and sizes, and only some so far had keys in the familiar QWERTY pattern.

The first commercial typewriting machine, the Sholes and Gilden typewriter was introduced by Remington (at the time principally manufacturing guns and sewing machines) in 1873. It had more than a passing resemblance to a sewing machine—not only had it the familiar sewing machine decoration of flowers on a black background, it was mounted on a treadle table just like a sewing machine, though here the operator moved the paper through the machine by pedaling. On these machines the bar with the letter on it was pushed up toward the paper, so the writing happened out of sight, within the mechanism—you could not see what you were typing.

According to a widely repeated myth, the QWERTY arrangement of keys was developed to slow typists down because the early Remington mechanism was unreliable and tended to jam if it were used too quickly. The story goes that Christopher Sholes, one of the inventors of Remington's typewriter, changed their original alphabetic layout to one where pairs of letters commonly used together are far away from each other. The theory was that the one-fingered typists of the day would take a while to hunt for the next letter, giving the type bar that had just been used time to fall out of the way of the next letter.

Quite apart from the fact that, if this were true, he didn't do a very good job—for instance, E and R, and N and G are in close proximity; this remains a delightful story that has very little evidence to support it. The Remington competed with a wide range of alternatives with different keyboard layouts and often won prizes for its speed of output. Later in life, the QWERTY keyboard would be challenged by a "scien-tifically designed" alternative, the Dvorak keyboard, which was sup-posedly much faster; yet much of the evidence for the Dvorak keyboard's benefits came from its developer. Detailed studies tended to show there was much more benefit from good training in using a QWERTY keyboard than from switching to Dvorak.

In fact, the explanation of the origins of the QWERTY keyboard is almost back to front. While it's true that Sholes was trying to separate frequently used letters, this arrangement wasn't intended to slow down the dumb one-fingered typist but to enable the typewriting machine to operate more quickly. Because the type bars that were likely to be used in sequence were well separated in the mechanism, they were less likely to interfere with each other, so could be operated at a higher speed. The QWERTY keyboard was not designed to slow down typists, but to enable them to speed up. When Muybridge encouraged Burk to use a typewriter, he was not only suggesting that he delegate part of the effort but that he make it more efficient, too.

Muybridge goes on to ask Burk to check whether a book that had been sent to him at the university from the Smithsonian museum had any animal locomotion pictures in it that might be of use in his new book (or books). He then goes on to complain about what was being offered by a "well known publisher in London." They would, it seems, only allow 25 percent royalties. To the modern author who is lucky to get half of this, 25 percent sounds very generous, but Muybridge wanted more. After all, he was going to be actively selling the book, acting as a salesman for the publisher, so why shouldn't he reap the rewards? He went on to break down the costs and compare them with the pricing. Rather than let the publisher make off with the profits, he suggested to Burk, couldn't the university publish it and in return for "an equitable commission"—perhaps 33 percent to 40 percent—Muybridge would sell the book and "attend to all matters connected therewith."

We don't have Burk's reply, but there is no evidence that the University of Pennsylvania ever took up Muybridge's offer. Eventually, in January 1898, he was to go down the route he said in the letter that he was "anxious to avoid" and signed up with the United Kingdom publisher Chapman and Hall for a percentage that probably made him wish he'd gone with his first offer—12 percent on the first book and 15 percent on the second.

The book wasn't Muybridge's only source of income. The lectures based on his Chicago shows proved a hit. He was charging himself out at the reasonable but hardly extortionate fee of 10 guineas each (around $50 then, roughly $1,000 now), including expenses to cover travel, an assistant, and the hire of the limelight that was used for the zoöpraxiscope as there was not yet a reliable source of electricity in many venues. Additional income came from sales of books and photographs at the events; Muybridge was doing very nicely. He had a uniformly successful time in 1895 and 1896, and was already planning his next series when he heard of problems in the United States, problems that would see him thrown back into a legal minefield.

∞∞

At the heart of the new lawsuit was his masterpiece, *Animal Locomotion*. The company responsible for the photographic part of the hybrid production, The Photogravure Company of New York, had threatened to sue the guarantors of the project to cover the expenditure on printing and storage of the copies of *Animal Locomotion* plates that had yet to be sold and were clogging up their warehouse.

Muybridge was horrified by this action, and wrote to Samuel Dickson, one of the Pennsylvania guarantors (Pepper had by this time left his formal position with the university), asserting that the printing company's action was a fraudulent attempt to gain money for no legally acceptable reason. Not only that, the Photogravure Company had said that, should they fail in their suit, they would destroy everything they held—negatives as well as prints. Not only would this be a terrible loss, it would cripple Muybridge's attempt to bring out a popular book based on his Pennsylvania work.

Muybridge had already had his suspicions about the Photogravure Company after earlier complaints from the company about the costs of the project. When he sailed for England he had written to Jesse Burk, asking him to forward any foreign or California letters but to open any others. He then went on:

> If any of them are from the Photo-Gravure Company of New York or on business connected with that company will you *immediately* send them to Mr. Edward H. Coates at the Academy of Fine Arts.

> I expect to have some legal disputes with that Company, and I will ask Mr. Coates to look after them in the interests of the guarantors and of myself.

> All other domestic letters please exercise your own judgement as to whether it is worthwhile to send on to me or to put them in the waste-paper-basket and I shall be content with your discretion.

Now that the Photogravure Company had struck, Muybridge had to act. In a series of frantic letters he gained the permission of the guarantors (who were probably wishing they had never been involved with a project that should by now have been long over) to travel to the United States and act on their behalf against the Photogravure Company. They agreed to pay his expenses. Muybridge was engaging in battle one last time.

By the next February (1896), Muybridge had taken a ship to Boston. It perhaps would have been more sensible to have been based in New York or Philadelphia, but Muybridge thought this a suitable location to be able to deal with the Photogravure Company and to be able to continue with his work on his new book. He met with the guarantors, in part to rally them to his cause, as some were inclined just to let the suit go by default and lose everything. Muybridge persuaded them that that would be damaging to the reputation of the university (and incidentally a problem for him).

While he was still rallying his own side, events were unfolding in New York that would put a new urgency behind the action. The Photogravure Company of New York filed for bankruptcy. Their assets, including all the material for *Animal Locomotion,* were seized and transferred into the hands of the Garfield Trust Company.

The most likely action for Muybridge to take at this point would be to try to persuade the guarantors to come forward with a little extra money in order to secure what was rightfully theirs. But chances are that by now many of them felt that they had done enough. In desperation Muybridge went back to the man who had started it all, William Pepper, living in retirement in Philadelphia.

A copy still exists of the telegram Muybridge sent to Pepper as problems mounted, given breathless energy by the convention of shortening telegraph messages to keep the costs down:

> Since writing saw opposing lawyer, sheriff removed plates, expenses running up. Can make no progress without cheque and authority.

Time was running out. Muybridge needed money, and quickly. Pepper wrote to the Pennsylvania university lawyer H. Galbraith Ward on June 15, 1896, ruefully commenting, "I suppose I shall never get over my foolish habit of ruining myself for the University interests." In fact, Pepper was to get a good deal out of the ensuing rescue, but he went on:

> I undertook my patronage of Muybridge, believing that it would promote the general recognition of the University. I believed also that it would lead to an important piece of useful work being done under scientific control. It has been expensive, but I think both results have been obtained. My associates do not seem to care to go further. I feel that it may help the University to be able to send presentation copies of these plates to important people from whom we seek concessions. Of course we shall never re-

ceive a dollar, and the advance, I suppose must be regarded as a further contribution.

Pepper was to pay $500. For this sum he would receive around 40 complete sets of *Animal Locomotion* and a thousand or two individual prints—around 33,000 photographs in all. Bearing in mind that they were still selling, admittedly slowly, at $1 each, this doesn't seem too bad a deal. However, Pepper wasn't out to get money for himself—the agreement was written in such a way that sales would first pay back the $500, then pay the remainder of the guarantors' original sums that had never been paid off, and finally, if anything were left over, it would go to Muybridge. Also Pepper was stretching himself at the time. In the same letter he comments:

> I am very heavily burdened now, because the Museums are growing splendidly and are consuming a great deal of money. I must give not less than $60,000 this year. This explains why I am careful about extra sums even of a small amount.

In earlier years Muybridge would have been more inclined to try to raise the money himself. It would be surprising if he had been unable to get together $500 to be paid off by a steady income from the photograph sales, plus the safe return of his precious negatives. His enthusiasm to have Pepper take on the debt reflected his new, settled status. It is difficult to avoid the supposition that he wanted to sort this out and get back to England, to writing his book and to the quiet life.

The one bone of contention between Muybridge and Pepper was the original negatives. Muybridge had no problem with Pepper holding on to them, but he needed access to them when required. Initially, and rather out of character, Pepper insisted that the agreement signed by Muybridge and the guarantors should read, "It is understood that the within includes a transfer to Dr. Pepper absolutely of the negatives as well as the plates [prints]."

Muybridge could not stand for this. He contacted Ward, the university lawyer, who sided with Muybridge and wrote to Pepper to inform him that as Muybridge was the photographer, he owned the negatives. But it was acceptable for Muybridge to sign over the right for Pepper to make as many prints from them as he required, provided he accepted that the original negatives belonged to Muybridge. This compromise was acceptable to all parties and it seemed at last that *Animal Locomotion* was saved. But there was one last hitch.

When Muybridge turned up to claim the copies of *Animal Locomotion*, the officers of the Garfield Trust Company would not release them. As far as they were concerned, these were assets of the bankrupt Photogravure Company of New York and no business of Mr. Muybridge. It took legal representation from H. Galbraith Ward to the receivers before Muybridge was allowed to arrange for 28 cases containing more than 33,000 prints and negatives, to be sent to Pepper. They must have made a serious dent in his attic space.

∽⊚∽

With the negatives secure, Muybridge could return to Kingston and his painstaking work on what had now become two books—*Animals in Motion* and *The Human Figure in Motion*. Published in 1899 and 1901, respectively, they provided a chance for a wider audience to appreciate just how much Muybridge had achieved. By now the halftone process had been developed, for the first time allowing photographs to be printed as part of a mass-produced book.

The two volumes were of a similar format—around 100 plates, showing perhaps 1,600 photographs, accompanied by text that described the history, the technology, and the lessons that had been learned from the experiments. From the research done for his many lectures, Muybridge was able to include comparisons of the photographs with many examples from art and could show how artists had failed—or succeeded—in capturing a true expression of frozen motion.

Reviews were enthusiastic. *Animals in Motion* generated real acclaim both for the book itself and the work that lay behind it. The *British Journal of Photography* (July 14, 1899) commented:

> The work of this eminent photographer was at one time very imperfectly appreciated and understood; but modern critics appear more disposed to rate it at its true value.

Similarly, *The Human Figure in Motion* went down well, though by now the commercial moving picture was familiar to newspaper readers, and so there were frequent references to the cinematographs and

bioscopes that were now almost taken for granted. Surprisingly, perhaps, given that this second book included a fair amount of the nude
work, there was little backlash against the potential for moral corrup-
tion. The reviewers did, however, warn that one should draw the line at
displaying the book too openly. So the *Graphic* of December 28, 1899,
rather obscurely informed the reader that:

> It will be understood that the volume, not being intended virginibus
> puerisque [literally: for virgins and for boys] (unless they be full-fledged
> students) should not be left on the drawing room table.

Despite these very specific concerns about *The Human Figure in Motion* ending up on drawing room tables, the books sold well, and had to
be reprinted several times.

After the production of the second book, Muybridge settled to a
quiet existence in the home of his cousin Catherine Smith. With Kate,
as she was known, and his friend George Lawrence, Miss Smith's half-
brother, he shared Park View, 2 Liverpool Road, a pleasant if unre-
markable detached house that still stands very much as it was in his
day, though now hidden from the road by huge metal gates. His intention to re-establish his roots in Kingston was made clear by his will,
dated March 14, 1904:

> Whereas I have resided over 40 years in the United States of America but
> have never become naturalized as an American citizen and I am now do-
> miciled in England with no intention of returning to reside permanently
> in America. . . .

He was not entirely unoccupied. The year 1904 found him engaged in the mildly eccentric activity of creating a scale model of the
Great Lakes in the garden of the Liverpool Road house. Quite why he
wanted to do this is not clear—the only part of his life he spent any
time near the Great Lakes was during the World's Columbian Exposition, which was sited on the shore of Lake Michigan, but that was an
adventure that had mixed memories for him. Even so, it could have
been that Muybridge had enjoyed early morning strolls along the
lakeside before the crowds appeared on the site, and pleasant memories of the location had stimulated him to start building his miniature
tribute to those dramatic pieces of water.

He was not to finish this, his last and least endeavor. Eadweard

Muybridge died on the May 8, 1904, a few weeks after his seventy-fourth birthday from a disease of the prostate, while digging out one of the lakes. He was cremated and his ashes were interred at Woking. This seems an odd choice of final resting place. Woking is, by road, nearly 20 miles away from Kingston. Muybridge's closest relative was his cousin Catherine, in whose house he was living, so why was he buried so far away?

At the time there were far fewer crematoria—Kingston did not get its own until the 1950s. The Brookwood crematorium opened at Woking in 1889, and was probably one of the few choices available. In his final markers at Brookwood, his adopted name was not to quite make it into the records. The crematorium register refers to "Edward James Muggeridge, known as Eudweard Muybridge," while the marker in the cemetery calls him Eadweard Maybridge.

Muybridge left his zoöpraxiscope and *Animal Locomotion* negatives to Kingston Borough, which now has the device on display in the museum along with a dramatic enlarged print of the San Francisco panorama. The rest of his photographic equipment went to his friend George Lawrence, while he bequeathed to his cousin Kate the interest from his money (capital amounting to £2,919 3s 7d) plus his books. On her death any remaining cash was to go to Kingston to contribute to the reference library. Muybridge seems not to have had anything to leave to the people who had been his closest friends while in America. But he had left a photographic and technical legacy to the world.

TWELVE

TIME LORD OR PATENT FRAUD?

The assignment was congenial to the unaggressive Muybridge, with its easy assurance of a comfortable berth and interesting associations. He was a picturesque person, with a great shock of hair and a long patriarchal beard, artistically stained with tobacco. In that distant time, nearly a half a century ago, when cigarettes were the essence of sin, Muybridge smoked them incessantly. He chronically wore tattered clothes and his hat was invariably full of holes.

Muybridge was a pictorial acquisition for the University of Pennsylvania in a double sense. He became one of the notables of the campus, where he photographed all manner of animals in motion, including the varsity track team.

From *A Million and One Nights*, Terry Ramsaye, 1926

After Muybridge died, his contribution to the origins of moving pictures was pushed into obscurity, in part due to the power of the Edison myth machine. The Lumière brothers were recognized as the originators of the commercial cinema, with Edison's kinetoscope considered the first working example.

There is no need to underplay the importance of the developments made by these other pioneers. The move from disc to celluloid was essential before a movie of any length could be shown (though ironically, the twenty-first century would see many people moving back to watching movies from a disc, in this case a DVD read by a laser). Yet the fact remained that Muybridge had shown the first moving pictures and had operated the first commercial cinema. His technology wasn't very good, but this is hardly surprising, considering it was the first.

Muybridge's technology was not the direct ancestor of modern cinematography, just as John Logie Baird's television with its crude me-chanical-optical scanner, was not from the same family tree as today's electronic technology. An even more dramatic parallel is with the work of Charles Babbage, universally recognized as the originator of the computer.

Babbage, born in 1791, inherited a small fortune from his banker father. Fond of extravagant social gatherings, he liked to have some-thing new and impressive to show off at his soirees. Initially, there was without doubt something of the dilettante about Babbage, though his intelligence and enthusiasm would push him into a very practical ap-plication of technology when faced with a tedious task. Through the summer of 1821, Babbage was drafted by his friend John Herschel to help with the painstaking job of checking an astronomical table.

Though the individual calculations to make up the numbers in the table were simple, there were so many to undertake that it made for a long and repetitive task. Babbage is reported as crying in desperation, "My God, Herschel! How I wish these calculations could be executed by steam!" Within a year he had devised a mechanical calculator, the difference engine, capable in principle of doing just that. The engine was a mechanical calculator. Initial values were set up on dials, a handle was cranked, and a complex series of gears was then worked through the calculation to produce a result on a series of numbered wheels.

Babbage would never build a complete difference engine (though one was built in the 1990s to demonstrate its practicality). Over the next 10 years he succeeded in getting 1/7th of the device assembled—enough to demonstrate its technical capabilities—before he was dis-tracted by a new and more enticing idea. His change of direction was not greeted with any enthusiasm by the British government, which had paid over £17,000 (around $1.7 million by current values) toward the development of the difference engine. The calculator would, after all, be valuable for everything from producing shipping tables to calculat-ing gunnery trajectories. But Babbage would never return to his origi-nal device.

The distraction came from a much grander idea—the analytical engine. This machine was inspired by the punched cards that had transformed the silk-weaving industry. Before a weaver's son Joseph-

Marie Jacquard introduced punched cards, it was not unusual for two skilled workers to produce just an inch of silk material in a day. After Jacquard's looms became popular, output rose to 2 feet a day. What's more, the result could be much more complex. One of Babbage's proud possessions, regularly showed off at his soirees, was a picture of Jacquard at work. At first glance this was a fine, detailed etching, but in fact it was a picture woven from silk with 24,000 rows of thread.

In the looms the punched cards—pieces of board with holes punched in, connected by small pieces of cloth to make a continuous train—controlled the different threads going into the loom, but Babbage realized that these cards consisted of information—instructions and data—that could be mechanically read and acted on. The same mechanism that could control a loom could power a calculating engine. But unlike the difference engine, this new analytical engine would not have its instructions built into the cogs of its mechanism. Instead, both the data to be worked on and the manipulations it underwent would be fed in on the punched cards. It would be a universal machine; in principle it seemed it could carry out any computational task.

Babbage would never construct the analytical engine; in fact, there is some doubt that his design could ever be built with the engineering tolerances of the day, but he did spend years exploring its theoretical working. He may have gone further if he had taken up the offer of help of a remarkable woman, Ada King. Ada, daughter of the poet Byron and married to William King, earl of Lovelace (hence the name she is often given of Ada Lovelace) was fascinated by Babbage and his work. In her teens there had been some talk of a marriage between Ada and Babbage, but Ada's mother was set on a more aristocratic match.

Ada translated a paper about the analytical engine from French into English, adding a long set of notes in which she described how the machine might be controlled. She also wrote to Babbage, offering her assistance, but he turned her down, reflecting the prevailing attitude toward women's intellectual capabilities at the time. So Babbage made one-seventh of his mechanical calculator and only theorized about a mechanical programmable computer. And there was no direct linkage from his work to the development of computers in the twentieth century.

Instead, the computer's family tree took a step back when Herman Hollerith in the United States started again from the punched card as a basis for his tabulators, the stock in trade of the Tabulating Machine Company, later renamed International Business Machines, and then simply IBM. In a practical sense Babbage's work was a dead end. But spiritually it was the turning point in the history of computers, earning Babbage the right to the title of father of the computer. Similarly, though Muybridge's technology would not be used in the true moving pictures, his developments put him in a similar paternal position.

Like computing, moving picture technology has seen a number of distinct phases. The basic technology of photography, though changing in sophistication as color was added and pictures took on a wider screen ratio, was in direct line of descent to that used by Muybridge all the way up to the 1990s. The big divide in those early years was the move from the disc, with its very limited playing time, to the celluloid film strip, which could go on as long as desired. Admittedly, for many years the movie would be projected from a number of reels lasting around 20 minutes, making it necessary for the projectionist to make the switch from projector to projector as seamlessly as possible, but the effect was of a limitless movie.

In the movie projectors introduced toward the end of the twentieth century even this limitation was removed—the reels are edited together onto a device called a platter that allows a whole movie to be shown without a change of projector—and thanks to a mechanism that feeds from the center of the reel, the movie can be shown again without rewinding. However, movie technology is currently undergoing a change that distances it even more from celluloid than the roll film shifted it away from Muybridge's discs—the transfer from analog to digital.

In the home we moved away from strips of film carrying a series of photographic images in the 1970s. The video camcorder killed celluloid for the home-movie makers. Commercial viewing, too, has shifted away not only from film but also from analog pictures. DVD players and digital TV broadcasts carry the picture and sound as a series of zeros and ones just as data are processed in a computer. And the same change has made its way to the movie theater. Projectors are finally available that can present digital images that look as good as film or even better. It is now possible for an image to be created, either directly with computer animation or using digital video cameras, stored and projected without ever passing through the stage of being a series of still pictures on a strip of film.

The advantages of the digital approach are huge. It escapes the dependence on chemical reactions that have always made it impossible to exert exact control over photography. Anything within an image can be changed, as becomes clear with the computer-modified movies that we accept as everyday now. The medium itself is less fragile. Movies no longer need to be shipped physically in cans around the country but can instead be dispatched electronically down a wire. Instead of the projectionist sitting in a large room at the hub of a series of noisy mechanical projectors, she can work at a computer and control the shows at many different multiplexes simultaneously. In the digital age the old cans of film are becoming as dated as the limelight used to illuminate the zoöpraxiscope discs. Muybridge's moving pictures had much more in common with twentieth-century movie technology than does the new digital equipment—even so, it all has the same conceptual origin.

Babbage retained his status from the first, but Muybridge went through a period of being sidelined. This tendency to remove Muybridge from the history of moving pictures began in his lifetime. In 1897, one Jules Fuerst gave a lecture on the development of cinema at the London Camera Club. He told his audience that it had been Edison who had first brought still pictures to life on the screen. Muybridge seems to have felt himself generally above such discussion, but in this case he was impelled to act. He wrote to the editor of the *Journal of the Camera Club*:

> If a recent lecture of the Camera Club was correctly reported in The Standard of 5th Nov., I have no doubt of one of the statements made by the lecturer causing you and some of the other members of your club considerable astonishment.

The statement in question claimed that the Edison kinetoscope was the first to synthesize movement, around 1893. Clearly Muybridge felt he was losing his place in history. He goes on to say:

During the last few years, numerous gentlemen in Europe and America have put forth claims to have been the first to demonstrate by synthesis the results of photographic analysis.

Having many years ago, practically retired from the field of photographic investigation, I have taken no part in this controversy. Since, however, the statement is gravely made to a body of scientific men assembled in your rooms, that an apparatus for showing "Animated Photographs"—so called—was not "invented" until about five years since, I thought it a not inappropriate occasion to send you, for the information of such members of your Club as care to take the trouble to read them, a few quotations in regard to some demonstrations of a similar character which were made so long ago that one may reasonably be excused for having forgotten about them.

The *Journal* printed both Muybridge's letter and a series of cuttings from 1878 to 1888 that described his work and the zoöpraxiscope. Mr. Fuerst was firmly put in his place.

That did not, however, stop a gradual move to sweep Muybridge's contribution under the carpet. Often this was without malice, simply a side effect of the Edison bandwagon. Since Newton's near sanctification, there had not been a man in the technical or scientific field who had received the same adulation as Edison, and only Einstein has had the same treatment since. However, there were exceptions to those sticking to the Edison party line. In 1915, H. C. Peterson, director of the Stanford University Museum, wrote an article on Muybridge's contribution:

He made possible the greatest aid to education that has ever been conceived. He created an entirely new and distinct industry. He made possible the bringing of the horrors of war and the blessings of peace to our eyes with such terrible force that we stand aghast at the brutality of the one and the failure of the other. We look and marvel. But to the wizard who created it all not twenty people of the fifteen million who daily witness the production of motion pictures on the screen have ever heard the name MUYBRIDGE.

Peterson was biased. In making Muybridge the wizard responsible for moving pictures, he was also boosting the reputation of California, of San Francisco, and of Stanford. ("Wizard" was an interesting choice of epithet; Edison was commonly referred to as "the wizard of Menlo Park.") Yet this bias doesn't take away from the fact that Muybridge's images projected in lifelike movement on the big screen were conceptual parents of cinema. Such facts would matter little to a man called Terry Ramsaye who burst on the scene in the 1920s. Ramsaye made an attack on Muybridge of such ferocity that his reputation has taken many decades to recover.

Ramsaye was the editor of a trade magazine for the cinema world, the *Motion Picture Herald*. In 1926, he brought out a lengthy (over 800 pages), gossipy history of the moving pictures called *A Million and One Nights*. With the fervor of a disciple, Ramsaye was to declare that Muybridge "had nothing to do with motion pictures at all." He felt that he was wiping away a myth, commenting:

> We have come to Muybridge and his tradition. We shall examine into [*sic*] the tale of a tale which by constant repetition has become the supreme classic reference of all motion picture history. It is the screen's accepted first chapter of Genesis, growing in authority and weight down the years.
>
> For at least twenty years every writer and every speaker on the annals of the motion picture has repeated with increasing assurance the time-worn story of the race horse pictures with which the late Eadweard Muybridge has been so orthodoxly [*sic*] credited with fathering the motion picture. Thereby the story has taken to itself the greatness of great names and the backing of high authority.
>
> But the supreme classic is supremely wrong.

The chapter Ramsaye dedicates to Muybridge is more than a simple misapprehension of the facts; it is a conscious, vituperative assault on the man's achievements and character. Ramsaye calls Muybridge "without energy or ambition" and clearly considers him an old hack with an over-inflated sense of his own importance. Not only does he not allow Muybridge any part in the development of moving pictures, but he even states that the zoöpraxiscope (a device that Ramsaye had never seen) had nothing to do with Muybridge, but had been invented by the French painter Meissonier after Muybridge's visit to France. This was despite it being demonstrated in the United States years before the French trip.

Muybridge expert Robert Bartlett Haas of the University of California at Los Angeles has since claimed that in that single chapter Ramsaye makes a good 50 incorrect statements that are provably false, from the inventor of the zoöpraxiscope to a plethora of inaccurate details, like misquoting Harry Larkyns's name as "Harry Larkyn" and describing Muybridge's arrival by boat at Calistoga. This would have been a particularly impressive feat, as Calistoga is 30 miles inland. Admittedly there was no evidence that Ramsaye did significant research, which accounts for some of the inaccuracy, but why was his account so blatantly biased? The driving force behind his battering of the Muybridge reputation was the influence of a bitter man: John D. Isaacs.

Isaacs has already appeared in this story, but his name could easily be forgotten—and this is the reason for the vitriol that comes through in Ramsaye's writing. Isaacs was the young engineer from the Southern Pacific Railroad who helped Muybridge develop the electromagnetic shutter used in many of the photographs. That same engineer who claimed during the lawsuit against Stanford that he, and not Muybridge, had devised the shutter.

When Ramsaye came to write his book, Isaacs was still alive, and it was from Isaacs that Ramsaye got much of his story, with a slant flavored by the bitterness of 40 years of being left out of the roll call of fame. Isaacs's claim was weak to say the least. The most he could have contributed was the idea of using an electromagnet to release the shutter. He did not work out the final design (this was done by another engineer, Paul Seiler of the California Electric Works). He had no connection with the early concept of motion photography, the original two-piece shutter, or the later development of the most significant part of Muybridge's contribution to motion pictures, the zoöpraxiscope. But Isaacs still felt the sleight strongly and transmitted his venom through Ramsaye's words.

Ramsaye's version of events would be sustained for at least two decades. In 1949, he wrote a letter to a Mr. Oscar N. Solbert, then the director of the Kodak photographic museum at Eastman House. Solbert had issued some comments on the development of motion pictures that gave Muybridge a role, and Ramsaye struck back with unabated distaste:

> The publicity release dated August 25th which came to my desk last week, finds your institution giving currency to certain aspects of the Muybridge myth. It is not surprising, of course, that it comes from Philadelphia, which is the home of the protagonists of the late Mr. Muybridge.

Note "the Muybridge myth"—Ramsaye was operating on the principle that mud sticks: Call something a myth often enough and people will start to believe it. He went on:

> The account of what Muybridge did is considerably erroneous. The mechanism which was used was evolved after Muybridge had failed, by John D. Isaacs, a technician and engineer in the service of Gov. Stanford's railroad interests. . . . Muybridge was not in fact a pioneer of instantaneous photography. John D. Isaacs was. There are records in the annals of the Southern Pacific Railway Company, and the whole story was set forth by me in my "A Million and One Nights"—the history of the motion picture, published in 1926. This is a work you will probably find fairly well known among the older members of the staff of the Eastman Kodak Company.

Sadly, Ramsaye's account seems to have been based solely on what the embittered Isaacs told him; no one has been able to find "records in the annals of the Southern Pacific Railway Company" that prove anything that supports his allegations. Despite the inaccuracy of *A Million and One Nights*, Ramsaye's account was an easy read, and it was not surprising that in later years, when a journalist or writer wanted to refer back to the earliest days of the motion pictures, Ramsaye's book would be used as a source. It was to darken Muybridge's name and contribution until more accurate histories were made widely available.

Looking back 100 years after Muybridge died, it is impossible not to be fascinated by this man. Eccentric, yes. Difficult, indubitably. Yet there was an indomitable side to his character that carried Muybridge through the murder trial and brought him out the other side even stronger. Something that drove him on to make a detailed study of motion. And something that brought him to operate his spectacular device at the World's Columbian Exposition.

In Muybridge, three desires fought for dominance. He was an inventor—of washing machines and clocks as well as photographic devices—and as such wanted to be admired for his originality and innovation. He had a deep desire to be accepted in the academic world—he would have loved to have had the professorship that the newspapers often unwittingly awarded him. Because of this desire, his unwarranted disgrace before the committee of the Royal Society hurt him immensely. Yet at the same time, he had a touch of the showman that encouraged him to promote his inventions to the public in the hope of making some money. Of the three, the urge to be taken seriously academically took precedence, but he could never forget the other two.

For Muybridge the ideal would be to bridge his desires, to combine academic status with public adulation and practical invention, just as photography fulfilled a Victorian ideal of combining art and science. His photographs were acclaimed for their beauty, but at the same time provided the opportunity to dissect and understand motion. This was science that could provide both benefits for the practical everyday world and help out the arts.

There is no doubt that Muybridge did make a contribution to artists' understanding of movement, though very quickly there was a backlash against a realist tendency to portray motion on canvas as it appeared in his photographs. The sculptor Auguste Rodin, while fascinated by insights provided by Muybridge's photographs, emphasized that the best art reflected reality as we see it, in motion, not the reality of the stop-frame photograph:

> It is the artist who is truthful and it is photography which lies, for in reality time does not stop, and if the artist succeeds in producing the impression of a movement which takes several moments for accomplishment, his work is certainly much less conventional than the scientific image, where time is abruptly suspended.

Rodin was right. Often the frozen postures of the Muybridge sequences, suspended in mid-stride, however technically accurate they may be, look unnatural. The skill of the artist is in representing movement in a still canvas, not faithfully echoing such snapshots. To be realistic the artist must paint what is not true, just as to give a realistic perspective view an artist has to make lines that are parallel in the real world converge. A perspective-view painting is an unreal representation; yet it accurately reflects how we see the world.

Though the photographs of stopped motion would not transform the art world, Muybridge's work at the University of Pennsylvania was important in providing a better understanding of the physiology of movement. And Muybridge did have an undoubted impact on the visual arts—the late twentieth-century artist Francis Bacon, for example, based several series of paintings directly on Muybridge images—yet the scientific and technical achievements of Muybridge's instantaneous photography and moving images were more significant than his reinterpretation of movement for the arts.

In the end it was neither photography alone nor the nature of motion that drove Muybridge but the synthesis of the two. Muybridge was drawn to the intersection of movement and the still moment that only photography would capture. And it was this interface, whether applied to stopping time or reassembling it into moving pictures, that will always define him as a unique contributor to the history of the modern media.

Muybridge was the Babbage of the moving picture, with the added bonus that he got his technology to work. There can be no doubt that Muybridge was the first to bring motion to the big screen and, most importantly, the man whose instantaneous photography of motion enabled us to stop time as never before.

SELECTED BIBLIOGRAPHY

Burns, E. Bradford. *Eadweard Muybridge in Guatemala, 1875: The Photographer as Social Recorder*. Los Angeles: University of California Press, 1987.

Carlton, Donna. *Looking for Little Egypt*. New York: International Dance Discovery, 1995.

Carroll, Lewis. *The Compete Works*. London: Reinhardt Books, 1999.

Clegg, Brian. *Light Years*. London: Piatkus, 2001.

Greeley, Horace. *An Overland Journey from New York to San Francisco in the Summer of 1859*. Lincoln: University of Nebraska Press, 1999.

Haas, Robert Bartlett. *Muybridge: Man in Motion*. Los Angeles: University of California Press, 1976.

Harris, David. *Eadweard Muybridge and the Photographic Panorama of San Francisco, 1850-1880*. New York: MIT Press, 1993.

Hendricks, Gordon. *Eadweard Muybridge: The Father of the Motion Picture*. New York: Grossman, 1975.

Herbert, Stephen. *Eadweard Muybridge: The Kingston Museum Bequest*. London: The Projection Box, 2004.

MacDonnell, Kevin. *Eadweard Muybridge: The Man Who Invented the Moving Picture*. London: Weidenfeld & Nicolson, 1973.

Mozley, Anita Ventura (Ed.). *Eadweard Muybridge: The Stanford Years 1872-1882*. Stanford: Stanford University, 1973.

Muybridge, Eadweard. *Animals in Motion*. New York: Dover, 2001.

Muybridge, Eadweard. *Complete Human and Animal Locomotion*. New York: Dover, 1979.

Muybridge, Eadweard. *The Human Figure in Motion*. New York: Dover, 1989.

Prodger, Phillip. *Time Stands Still: Muybridge and the Instantaneous Photography Movement*. New York: Oxford University Press, 2003.

Ramsaye, Terry. *A Million and One Nights*. London: Frank Cass, 1964.

Solnit, Rebecca. *River of Shadows: Eadweard Muybridge and the Technological Wild West*. New York: Viking, 2003.

Tutorow, Norman E. *The Governor: The Life and Legacy of Leland Stanford, a California Colossus*. Norman, OK: Arthur H. Clark, 2004.

Newspapers and magazines from originals and copies held by the publishers, the British Library, and Kingston Local History Room.

Papers and letters from copies and originals held by Kingston Local History Room.

ACKNOWLEDGMENTS

Many thanks to those who have helped me bring this book to life. Particularly to my agent, Peter Cox, and to Andrew Gillman for cinematic inspiration, and Jeff Robbins, an editor who really understands science and the fascination of the Muybridge story.

Also to the people who have helped with research, including Jill Lamb and Emma Rummins in Kingston-upon-Thames Local History section, Ged Doonan at Leeds Public Library, and Mrs. Joanna Corden of the Royal Society. And to those for whom the Muybridge story has been an inspiration in the past, including the relentless Janet Leigh, daughter of the defense lawyer at Muybridge's murder trial, Wirt Pendegast, and academics Robert Haas and Gordon Hendricks.

Credits:

INDEX